The Poultry Raiser's Guide
Our Money Makers

by Kit Clover

with an introduction by Jackson Chambers

This work contains material that was originally published in 1885.

This publication is within the Public Domain.

*This edition is reprinted for educational purposes
and in accordance with all applicable Federal Laws.*

Introduction Copyright 2017 by Jackson Chambers

IMPORTANT NOTE & DISCLAIMER

PLEASE NOTE :

As with all reprinted books of this age that are intended to perfectly reproduce the original edition, considerable pains and effort had to be undertaken to correct fading and sometimes outright damage to existing proofs of this title.

Some books require total rebuilding of some pages from multiple different digital proofs, which can be painstaking and time consuming. Despite this, imperfections still sometimes exist in the final proof and may detract slightly from the visual appearance of the text.

This book appears exactly as it did when it was first printed.

DISCLAIMER :

Due to the age of this book, some methods or practices may have been deemed unsafe or unacceptable in the interim years.

In utilizing the information herein, you do so at your own risk. We republish antiquarian books without judgment or revisionism, solely for their historical and cultural importance, and for educational purposes.

Please use due diligence when gifting any antiquated book, especially to a minor.

Self Reliance Books

Get more historic titles on animal and stock breeding, gardening and old fashioned skills by visiting us at:

http://selfreliancebooks.blogspot.com/

INTRODUCTION

I am very pleased to present to you another essential poultry title – *The Poultry Raiser's Guide*. It was written by *Kit Clover*, and first published in 1885, making it over one-hundred-and-thirty years old.

With so many people embarking on self-sustainability and self-reliance endeavors these days, we like to publish lots of poultry titles. These are are our best-selling books.

The rise in the cost of eggs and meat has been accompanied by a decline in their quality, and once you taste your own home-raised eggs and meat, you will *never* want to eat either one from a store ever again!

People are more conscious than ever about where the food they put on the dining table for their family to eat has come from – we consider everything from the ethical treatment of the animals, what they are fed, over-vaccination, the over-use of antibiotics, the quality of the meat, and the hygiene of slaughter houses. We don't know the answer to *any* of the questions we may have about where our meat and eggs come from – unless we raise them ourselves.

The Poultry Raiser's Guide features chapters on *Poultry Houses and Yards, Plans for Houses, Hatching and Caring for Young Chicks, What to Feed, Incubators, Brooders, Fattening Fowls* and lots more, and it also covers raising and breeding Chickens, Ducks, Geese, Pea Fowl, Turkeys, Guinea Fowl, Pheasants, Swans, and Pigeons.

This valuable old text should be in the library of of all poultry farmers, be they back-yard egg producers or commercial breeders. It also makes for a great starting point for a beginner, or somebody contemplating taking the plunge into back-yard or full-scale poultry farming too.

Jackson Chambers,
State of Jefferson, December 2017

INTRODUCTORY.

When I had fully decided to turn a part of my possessions in Cloverdale into a poultry farm, I, wisely as I thought, concluded that I would raise poultry *books* first, and poultry afterward. I did so. There's a box full of those books up in the attic at this very minute. And what did they tell me? Well, one said I must surely buy the author's recipes for all poultry remedies, if I hoped to succeed in raising chickens. I bought the recipes. They are up there in that identical starch box, also. As this book treated only of diseases, I sent for another. I wished to know what breed of fowls to raise, and the new book told me. Gave cuts of all sort and kinds of fowls; colored plates of Black Spanish that were all red and green, and Brown Leghorns with red hackles and green tails. And also said the author had the very finest strains of—well, of several breeds advertised for sale at a good round price. I ordered eggs. I wanted to know something of incubators, so sent for a book. It gave glowing accounts of eggs hatched in incubators, more especially the one patented by the author, price so much. I bought an incubator. I had already subscribed for a half dozen poultry papers, and seeing an advertisement, "PLAN FOR A BROODER FREE, on receipt of ten cents to pay postage," I forwarded ten cents. Received a circular in which the inventor offered to send me plans and specifications complete for the small sum of one dollar. I must have a brooder, so I sent for plans. When I had paid my carpenter for his time, and bought the truck "specified," I had a brooder of no special account, and by no means a cheap one either. Then I *must* have a chicken house, so sent for book on that subject, and when received, did not find a plan in it that suited my purpose. Saw an advertisement that read: "FREE. Receipt for Preserving Eggs." Sent for it. Got a circular. Could have the receipt for two dollars. Of course I wanted to preserve eggs, so bought the receipt, and several more. Some good and some not so good, and some—well, we will draw a vail over that subject. I bought roup pills and cholera medicines and egg preservatives, all before my incubator had begun to get in its work; because, you see, the books told me I must have such things, and I wanted to be on time. I have not used those remedies yet, but I may need them sometimes. One book told me to use sulphur and lard on the heads of young chicks, to prevent the gapes; and another told me *not* to do so. One advised the use of incubators, and one said "stick to your old hens." Just how I succeeded in the chicken business, is not necessary for me to state here, suffice it to say I have no incubator to introduce, no plans for brooders, no fowls or eggs to advertise by means of this book, and no roup pills to sell. I simply give what I believe to be the information needed by poultry raisers, as nearly correct as my experience enables me to do; and because the subject is so voluminous, I will not sandwich in any "personal experience," but condense the most information in the smallest space, that the cost of printing may not put the book out of the reach of any.

KIT CLOVER.

The Poultry Raiser's Guide.

POULTRY HOUSES AND YARDS.

In considering the subject of poultry houses and yards, I have taken plans and specifications just wherever I happened to find those that suited me; as "hen houses" do not bear the mark, "pat. applied for," I shall not stop to give authority.

A poultry house should be warm in winter and cool in summer, dry, well ventilated and sunny; or to make a short story long, the poultry raiser should first,

Select for the site of the poultry house and yard a dry soil. Dampness causes or intensifies that scourge of poultry, the roup; it renders cleanliness next to impossible, and is indirectly the fruitful mother of a variety of diseases. If the soil is not naturally dry, drain it, and make it as dry as possible. Then do not commit the too common error of setting your house so low that the first rain will cause a miniature flood, and make the inside of your fowl house resemble a duck pond. Set your house above the natural level of the soil and fill up to it, so that the land will slope from it each way, and form a good water-shed. Dry earth used within the house, scattered over the floors, helps to render the atmosphere dry, besides, being an admirable absorbent of those gases which are a valuable component part of fertilizers, but deadly to your stock. **Dry earth is the best absorbent, but sawdust, haychaff, or dried leaves are all better than nothing.**

Secure sunlight. Let your fowl house face the south, with southern and eastern windows. You can not overestimate the value of sunlight for your fowls. Sunlight with its bright lances, will put to flight that dire army led by roup, whose *aides de camp* are colds, catarrhs, rheumatic affections, diarrhea and cholera.

Let one ray of sunlight fall across your hen house, and every fowl in it will find the sunlight, and struggle for a place to nestle down in it.

Don't forget fresh air, or in other words, provide suitable ventilation. A direct draft should be avoided at all times, but fresh air and means for the escape of foul gases must be provided. The want of fresh air leads to a weakened state of the constitution, the blood fails to be properly aerated and becomes thick, dark and sluggish; hens cease to lay, contagious and epidemic diseases break out among them, and the loss of a part, or the whole of the flock ensues. And all this because fresh air was not provided!

Cleanliness is necessary. Buildings should be so arranged as to be easily cleaned. Dry earth should be provided for the floors, it being one of the best deodorizers known. Filth produces vermin, vermin produces disease, and disease produces death.

Provide shade in the yards and runs. If you doubt the need of this, stand for a half hour without hat or other protection, under the blazing rays of a July or August sun, and after this experience, if you do not die from a sunstroke, think how your fowls would enjoy protection from the vertical rays of "the too-near-approaching sun." Trees are best, but boxes and boards may be propped up, and will afford a satisfactory substitute.

Pure water. A liberal allowance of fresh, pure water of a moderate temperature is a thing absolutely necessary to the health of any kind of fowls. Through the summer months all that is needed is a receptacle that can not be fouled, and will not slop over and make trouble, or be scratched full of the loose dirt of the floor wherever a fowl scratches or wallows. A very good drinking trough is made of tin, about six inches high and a foot long, and placed on the side of the house about eight inches above the floor. In such a tank, water supplied in the morning will stay sweet all day during the months when no water congeals; but when winter puts in its appearance, we must have recourse to a better form.

Oyster cans that have been opened on the side will serve the purpose if one studies economy, provided the rough edges of the tin are carefully hammered down, so as not to injure the fowls comb or wattles.

Convenience should be sought after. "Time is money;" therefore, build so as to save time. A few dollars more spent in making a house convenient is money well invested; it will pay large dividends in actual gain, not only of time, but money itself. For, if things are convenient to clean they will be kept clean, and cleanliness is, as we know, absolutely essential to success.

Whitewash the poultry houses thoroughly with a mixture of two tablespoonsful carbolic acid, or a pound of sulphur to a pail of wash.

PLANS.

The cheapest plan for a hen house, is one made of sod, and thatched with straw, with glass door and window. An old barn can also be converted into a poultry house, by putting windows in the south side, and papering the inner walls with tar paper, or even common newspaper, provided a little carbolic acid is mixed with the paste.

Twelve hens and one cock are all the fowls that should be kept in one house, as hens in large flocks do not do as well as in small numbers. A moderate sized range is better than one too large, or too small. If the range is too large the fowls rove about and wear off their flesh, while they trample down and soil the entire territory. If too small it breeds vermin and disease, unless great care is taken. 15x25 is small enough for a flock of ten fowls.

A cheap and convenient poultry house, to accommodate from twelve to twenty fowls, suitable for one or two breeds, may be built as follows: Fig. 1 represents the ground plan of the house, which is 16 feet long by 8 feet wide, making two rooms for the hens, one 6x8 and the other 8x8. An entry or hall 2x8 runs along the building, so that doors communicate with both rooms. This hall can be used for the storing of grain, and nest boxes can be arranged along it, so that the eggs can be gathered without entering the room where the fowls are kept: A A represent the small doors through which the fowls pass into their yards. B B represent roosts, and D D D the doors of the building. Beneath the roosts a board to catch droppings may be placed. The floor is to be boarded or to be of dry earth, as the builder may desire.

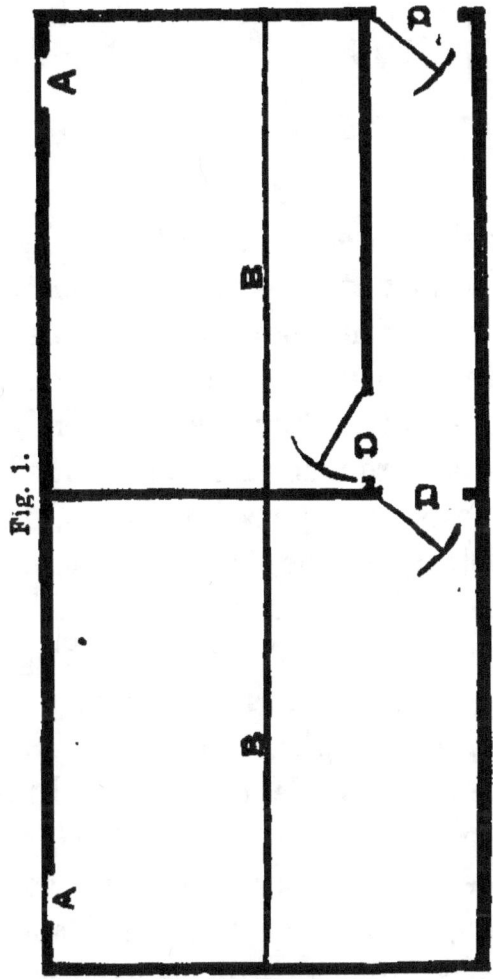

In fig. 2 we give a plan that can be built for a few dollars and unites cleanliness and convenience, and it can be built on any scale desired, to suit the space at command.

Figure 2 shows the front elevation. The plan of floor is simple. The house should face the south or southeast. A passage three and one-half feet wide runs

Fig. 2.

along the back, the entire length of the house, separated from the rest of the space by a partition of three boards high, and lath above that to the ceiling. The entrance to the house is by means of a door at each end of the passage way. Four doors open from the passage into as many rooms, each partitioned like the passage. The nests, are placed in tiers, two high; the lower tier resting on the floor and facing the passage partition, the middle board of which is hinged, and raises up, thus allowing the eggs to be gathered from the passage. The perches are strong frames, being ing about 20 inches high; the cross-bars being 18 inches apart. These are movable, and can be lifted out of the way when cleaning. The windows in front are 8 feet high and 5 feet wide, being made in two large sashes, the upper hinged at the middle and opening inward. Along the back are a row of ventilators, placed close under the eaves, which, when the windows are dropped back about 12 to 18 inches, give perfect ventilation, and yet no draft on the fowls.

The end windows are single sashes, and immovable. The holes for egress and ingress of the fowls are shown in Fig. 2. The height of the building is $6\frac{1}{4}$ on the back, and 10 feet in front.

The following illustration represents a very neat house. It is 10x12 feet; 6½ feet in front, and 4 feet in

the rear peak 8 feet. B B are shutters to fasten closely at night. Door for egress of fowls at D.

We regard Figure 5 as the best and most economical poultry house that can be built. It is economical because it gives the greatest amount of ground floor to the smallest possible amount of roofing. A building 8 feet square will give nearly 12x8 feet on the ground, in consequence of the addition of the windows.

Fig. 5.

A cheap, durable, quickly built poultry house (Fig. 6) of any design or size can be constructed in the following manner, using Three-ply Standard Roofing (tarred felt) for weather boarding as well as roofing. Build the frame in the ordinary manner, place the stud-

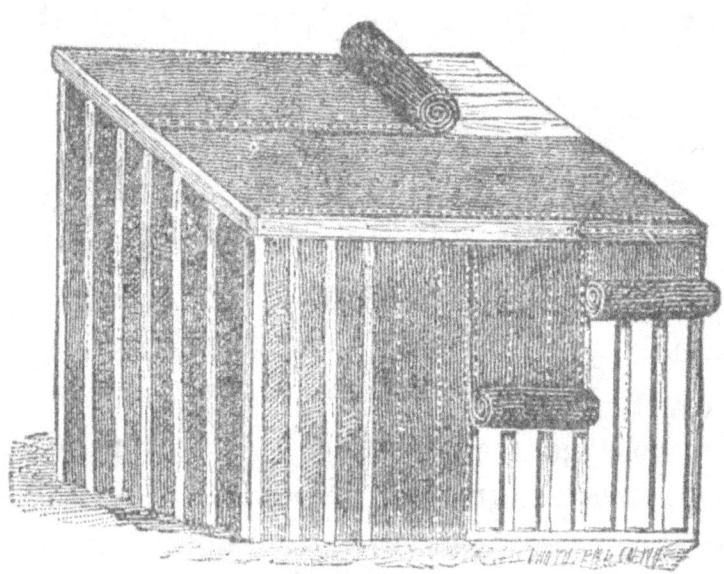

Fig. 6.

ding (any kind of lumber, old, or new will answer in place of studding) eight inches apart from center to center. Nail on the three-ply perpendicularly (as shown in the cut) with large carpet tacks, about every three inches. Do not lap it, but bring it edge to edge, and nail battens on the studding over it. When finished apply two coats of Standard Cement. After the coating is dry the house may be whitewashed, if desired. Three gallons of Standard Cement covers one hundred square feet—two coats.

In building on the above plan, it will be readily seen that no skilled workmen are needed and that the three-ply serves the double purpose of siding and lining, thus saving considerable in labor and material.

Fig. 7.

Figure 7 represents a fancier's house with yards adjoining.

EARLY CHICKEN HOUSE.

There is no use in trying to raise chickens in February and early March certainly, without suitable accommodations. We here present a style of house for early chickens. The three essentials of a glazed front to the south, thorough ventilation, and apparatus for artificial warmth, are all secured in this plan. The outside door being opened in the figure, a range of shelves is shown, which, it will be observed, do not occupy the whole room, but there is a passage-way of the width of the door, running the length of the building. The room narrows as it approaches the top, therefore the partitions upon the shelves are placed further apart the higher the latter are, in order to give sufficient room in each apartment, which is for a hen with her brood. The shelves are 18 inches apart in the clear. The partitions are of wire cloth, fine mesh. The small doors next the passage, one of which communicates with each apart-

EARLY CHICKEN HOUSE.

ment, are removed in the illustration, in order to show the shelves and partitions. A stove in the lean-to part is for the purpose of making a slight fire (using wood) towards night, and continuing it until bed time, and starting it again for a little while in the morning. Generally the glass exposure will secure abundant warmth in the day-time, especially as spring approaches. A stove is placed in a pit or cellar properly drained, and walled with masonary. The pipe is close to the floor, and runs the whole length of the building.

The building is 8 feet high at the peak, and of the same width in length at the foundation, and the length must be proportioned to the extent of the business.

We give in figure 11 a very good style of movable roost, although perhaps more elaborate and expensive than farmers or small poultry raisers would care to adopt.

Fig. 11.

Chicken coops are made of all shapes and sizes, but for general use, expense being taken into consideration, nothing better than the old-fashioned "A" coop has been designed. It is well to have a floor inside to raise the chicks above the wet earth--an important matter when a cold storm is in progress. An A-shaped lath run in front of the coop to enable the hen to get at the earth and into the sun, is desirable. It will be found convenient to have the bottom board of the back of the coop hung on hinges, so that the breeder can easily reach the hen and chicks whenever he wishes to do so.

A barrel laid down upon its side and propped up, so as to prevent water from standing in it in case of a

Fig. 12.

driving storm, will answer very well for a coop, though it is not convenient to get at the hen or chickens when necessary to examine them.

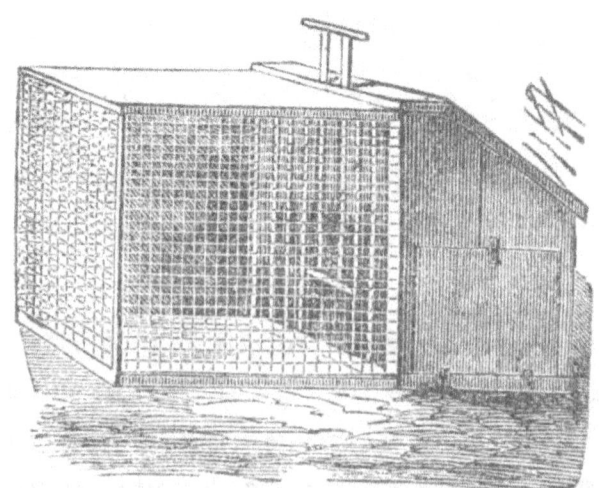

Cat-proof, rat-proof, and hawk-proof coops and runs are demanded in some places. They are not difficult to make, but fine-meshed wire enters into their construction. Fig. 1 gives a very good idea of such a coop.

YARDS AND FENCES.

Next comes the yard. If you keep but one variety, and can allow them to run at large, no yard is needed. But if you have more than one breed, which you wish to keep from intermixing, you must build a yard, and the questions of size, height of fence, fencing materials and other details must be settled.

Make your yard as large as you can afford to. The larger the yard the better your fowls will thrive. A yard 20x50 feet will answer for a flock of twenty fowls, and you can keep that number in one-half the space, but they will not do so well if thus crowded. The height of your fence will depend upon the breed kept. Hamburgh, Leghorns and Games require a fence not less than six feet in height, while eight feet is still better. The Asiatics can be kept in a yard which would hold a pig. A fence made of one length of lath and pointed on top will keep Plymouth Rocks safely. You can build either permanent or movable fences. The latter possess some very decided advantages, as fresh soil for fowls is very desirable. Movable fences are built in the form of the old Virginia snake fence, by allowing the rails to project a little beyond the pickets. They can also be made in lengths, as shown in the following figures.

Fig. 1 shows a length of fence with the posts. Fig. 2 shows a sharpened post, with the two hooks upon which the fence rails are hung. The length of the post should depend upon the character of the soil into which it is to be driven.

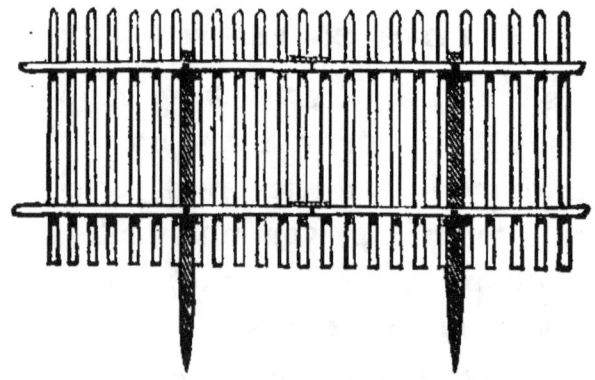

Fig. 1.

A very good permanent fence can be built by setting posts at the usual distance, providing light top and bottom rails, and nailing around the bottom next to the earth a board, above which laths sharpened at one end are used for pickets. This is an economical fence. The best fencing material is the galvanized wire netting, which can be procured of almost any desired width and at very reasonable rates. No fence rails are needed when wire netting is used, but a good, wide bottom board should always be provided.

Fig. 2.

Here is the picture of a fence with a revolving top, (patent applied for by some one, we forget who); also three stretches of wire drawn tightly above the tops of

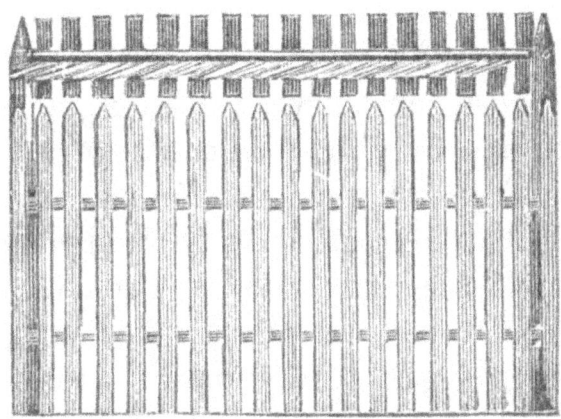

Fig. 1.

a moderately high picket fence, are a very good preevntative of high flying. The wires should be about six inches apart, one above the other.

I think now that I have given suggestions enough regarding "houses, coops and yards" to set any one up in the business, provided he has lumber, nails, and a good claw hammer, and I will "draw my remarks to a close" by adding, be sure to keep houses, fences and coops *well* whitewashed, with the carbolic acid or sulphur addition.

And now that we have a place to keep our chickens, we will proceed to populate them, which must be done by purchasing hens, or incubators and eggs. As the two are entirely different in management, we will consider the old hen question first, with the supposition that she is a desirable breed, and that her eggs are saved for hatching. And here is a good place to say, make the floor of your chicken houses of dry earth if possible, and place the nests of sitting hens on the earth, or if boxes are used, fill in two or three inches of earth. Poultry raisers can not be reminded too often of this.

HATCHING AND CARING FOR YONUG CHICKS.

Eggs designed for hatching should be collected as soon as laid, especially so when the weather is cold, or the winds sharp. They should be laid singly on soft material. A flat box containing a single layer of eggs packed carefully in bran, with top fastened on, so that the box can be turned every day, is the best and easiest way of packing eggs. A moderately damp cellar is the best place to keep them at all seasons of the year. It makes no difference how the eggs are laid in the box, if care is taken to turn them daily. Eggs if carefully packed for shipping will hatch well even if carried long distances, and will keep from three to five weeks.

Fresh eggs hatch quicker than old ones, and make stronger chickens.

Any of the Asiatic breeds make good mothers, as they are naturally tame and gentle; but heavy hens must not be set on the eggs of Hamburgs or any of the thin shelled varieties, because of their liability to break the eggs.

In raising chickens one should remember that the early hatchers always sell for the most money; and if one intends to raise poultry for market, they must observe this fact. All things considered, I think it is the better plan to make an incubator do the early hatching, for when eggs sell for thirty-five cents a dozen, it does not pay to have a hen spend her time over eight or ten chicks. But if one decides to set hens, then a few hints as to the forcing process will be useful. By the first of January provide good, warm nests, put dry earth, sand or ashes in the box, and cut straw over that, and make sure that there is no possible chance for a draught to strike the nest. For feed, give mixed feed warmed; or to be explicit, feed middlings mixed with warm sweet milk, in which is mixed a little pepper, for morning; wheat and chess, or mill screenings at noon, and parched corn at night. Feed everything comfortably hot. Give all the pounded bone and oyster shell that they will eat. If bone meal is used, roast it, and mix plenty of it in their milk feed in the morning. Supply plenty of sand, and do not neglect to provide a large dust box.

As soon as a hen begins to sit, she should be removed to some quiet corner where other hens will not trouble her by trying to lay in the same nest. A hatching house is a desirable thing to have, even if improvised by a board partition set up across one end of the hen house. Provided with dust boxes, drinking cups and plenty of corn, the hens will need no further care until the chicks are hatched.

TO MAKE ARTIFICIAL HEAT.

Take a box at least sixteen inches square, twenty inches is a better size, and fourteen inches high; throw in fresh horse-stable manure four inches deep; when pressed firmly down on the top of this, spread an inch of dry earth, and finish with straw, cut, if you have it. Heat the nest with a hot stove-lid or flat piece of rock, warm the eggs in milk-warm water, set your hen at night and keep her nest very dark for the first three days. A hen at all inclined to set, will get down to work in earnest under such treatment.

All eggs set early in the season must be sprinkled three times a week, or in fact, must be so treated at any season, if the hens are confined. If the nests are placed on the ground, and the hens come off for their food in the morning while the grass is wet with dew, they will accumulate sufficient moisture on their feathers, but not otherwise.

Should the nest become fouled by the breaking of an egg, take the hen off carefully, feed and water her near the dust box, wash the eggs in blood-warm water, put clean straw in the nest, and let the hen return to it. Eggs will not hatch if the shell is encrusted with broken egg, as the chicks can not get out of the shell.

It is best to disturb the hen as little as possible while sitting, and it is almost useless to attempt to help chicks out of the shell. If the hen gets restless when the first few chicks come out, and is inclined to leave the nest, take the chicks away from her and put in a brooder. If she refuses to own the chicks when taken back to her, it is best to take the entire brood from her, as some hens will kill all such chickens. On the other hand a home-made brooder will raise chickens with as little care and expense as will the mother hen, while the hen may be set a second time, or turned out to range, and be on hand as an egg producer again.

CARE OF CHICKS.

Just as soon as the chick can lift its head up in the downy feathers of the hen, it becomes exposed to lice. The hen and chicks should be lifted out, the nest changed for clean straw, the hen replaced, and the head and a little of the back of each chick rubbed with a mixture of lard and sulphur, with a trifle of carbolic acid, or even coal oil mixed with it. But great care must be used that the mixture does not get into the chickens' eyes. Where leaf tobacco is grown, some steep a little tobacco in the lard. Possibly, prepared tobacco might be good if a smaller quantity was used. Two ounces of leaf tobacco to a pint of lard is considered about the right proportion; then put the chicks back and let them remain as quiet as possible for at least twenty-four hours. The chicks do not need any food before that time, as a chick absorbs the yolk of the egg just before coming out of the shell, and all they need is brooding, receiving much more strength by that, than in any other possible way.

WHAT TO FEED.

Stale bread dampened in hot milk or water is the best feed for the first week a trifle of hard boiled egg, or raw onion chopped fine and mixed with the bread is an improvement, as it supplies both the animal and vegetable food needed. Middlings mixed with equal parts Indian meal moistened with *scalding* milk or water, is also good if the flock is large, and the bread crumbs do not hold out, and the food should be given once every hour or two if possible. As the chicks grow, cracked wheat is the very best food for them; and in time, wheat chess, mill screenings, and finally cracked corn may be fed to them, and may be kept constantly by them if the other poultry is excluded from their yards. Be ca~~~~~~~~~~~~~
corn meal wet with cold water, as i
and is in every way the very poo

chickens. Give water in shallow platters, to which add a bit of lime, and keep the drinking water at all times clean and fresh.

COOPS.

One must use his judgment as to what is needed in the coops. In wet weather a board floor is a necessity, if cold and damp; *cut* straw or hay chaff is warm and comfortable; never use hay or straw uncut, as the chicks become tangled and the old hen steps on them; but in dry, hot weather, set the coop on freshly spaded earth, and move the coop to a freshly spaded spot, if convenient, once in three or four days. All fowls should have a patch of fresh earth spaded over for them every day if possible. If one doubts this assertion, let him spade and plant a choice flower or vegetable bed, and if there are any fowls within a mile of the place, he will be sure to find every last one of them wallowing in it within the next half hour; which proves not the perverseness of all chicken nature, but the fact that they delight in freshly turned soil.

In spring hatches it is a good plan to have your chickens out about the time the grass starts, as they prefer to get at least one half of their living in this way; besides, as the grass gets higher they get badly drabbled with dew if allowed to run at large, and cold and wet are absolutely death to chickens.

WIND BREAKS.

Provide your chickens with wind breaks. Set the ——————— high, dry sheltered place, with a southern ———, and on the sunny side of the coop ——— 'h strong sticks driven into the ground in such a position as to give the chicks a warm sunny place to congregate "and consider the needs of the coming generation." If chicks hatch early a very good "yard" may be made by nailing four boards together and stretching unbleached muslin over the top. Spaces should be cut in one side to admit the chicks, then spade up a piece of ground at the front of the coop, rake in wheat, place the "yard" over it, and let in the chicks. They will have a lively time scratching out the wheat, and the muslin, while allowing the warmth of the sun to penetrate the yard, completely secludes the wind.

DO INCUBATORS PAY?

An article in *The Poultry Keeper* sums the matter up as follows: "The above is a natural inquiry. No one desires to purchase an incubator unless satisfied that it *will* pay. An incubator, like anything else, pays in *proportion* to the *manner* in which it is operated. It should not be considered as a toy, a novelty, or something to satisfy those who are curious, but a *practical machine* for doing effective work. The majority of those who purchase incubators are not very liberal in their demands regarding what their incubators should do, and because the incubators are expected to hatch the eggs placed in them, do not consider that comparisons may be made with results when hens are used. The true way of noticing the value of an incubator depends upon a great many circumstances. Because it may be self-regulating does not imply that the regulator is human. It has no brains, and can only perform its duty as the manager may desire. No farmer expects his stock to be profitable without the exercise of the best care and judgment, and he should not expect an incubator to do *more* than is required of the average flocks of poultry, or from other investments. Hatching with an incubator is simply a *wholesale* method of raising poultry, and, in *proportion* to capital invested, is more profitable than any other system."

VALUE OF EARLY CHICKS.

The early chicks that get into market in January, February, March and April, are known as "broilers," and sell best when not over a pound in weight. The price is then anywhere from 60 cents to $1 per pound. Sometimes even three-quarters of a pound is a better weight than a pound. They seem to remain at that uniform price for the *entire* chick until grown fowls, reach the market. That is, they sell at about 75 cents per pound when a pound in weight. Then afterwards, along in May, the preferred weight is one and a-half pounds, the price being usually about 50 cents per pound, or 75 cents for the chick. As June approaches, those of two pounds weight become more salable, the price being in the neighborhood of $37\frac{1}{2}$ cents per pound, the 75 cents per chick still being maintained. And so the figures and weights keep pace, in contrary directions, until late in the fall, when the grown fowl of eight or nine pounds goes to market and bring about 75 cents. Of course we may not be very exact in regard to the figures, as locations, markets and other causes may vary them in either direction, but it requires only a glance to show that the profit is in the *early* chicks, and as we cannot procure early chicks, owing to human ingenuity not as yet being sufficient to *make* a hen sit until she is inclined to do so, the *true value* of an incubator can only be estimated when we consider that it enables us to *hatch at any time of the year*.

CAPITAL REQUIRED.

If we estimate the cost of hatching a certain number of chicks under hens, we must allow the usual time of three weeks for incubation, and, in cold weather, nine weeks at least for the hens to carry the chicks. This is a loss of twelve weeks of the time on the part of the hens, as, by "breaking" them from sitting, they would probably lay at least eighteen eggs each, which, at winter prices for eggs, would be fifty cents. We make a low estimate, as we wish to be within the bonds of fairness in our calculations. If fifty chicks are hatched under five hens (ten chicks each, which is not usual in winter, the majority of them hatching fewer numbers), the cost for loss of eggs is $2.50, which may be charged to care, but the poultryman must *still care* for them, as he has to look after the hens as well as the chicks. His time must still be devoted to them, for winter is a precarious season with chicks under hens, or else we would not secure such high prices. Then, there is the value of the hens themselves, and the food they consume along with the chicks. The incubator, on the contrary, if it hatches fifty chicks, is ready to begin again, and while the five hens are hatching and raising fifty chicks, will, in twelve weeks bring out four broods, or 200 chicks. Just there we wish our friends to look well, for we make this comparison: "That while a hen is *hatching* and *providing* for her brood, *for every chick she hatches and raises the incubator hatches four.*" We did not mention that the hen may trample some of them to death, and say everything we could against her, for that would not be fair, for we may also lose some in the brooders, and prefer not to be ungenerous to the hen.

A LOW ESTIMATE.

We have known incubators to hatch 88 per cent. of the *gross number*. We do not like the method of guaranteeing a machine to hatch 90 per cent. of *fertile eggs*. Such a plan of explanation is unfair. An incubator may contain 100 eggs, and ninety of them may be infertile. From the other ten eggs nine chicks may be hatched, and as it is 90 per cent. of the fertile eggs, the idea is conveyed that the incubator hatched ninety chicks when only nine was the result. When we put eggs under a hen, we say we secured ten chicks from thirteen eggs, the *gross* number. That is the only fair way of estimating, for an infertile egg costs as much as one that hatches. But suppose we take twelve weeks, the time a hen hatches and leaves her brood, and see what an incubator holding one hundred eggs will do. We put in one hundred eggs and secure fifty chicks, or 50 per cent. *gross* number. We will say that we were unlucky, and secured forty chicks and lost fifteen of them. Here we have only twenty-five chicks left for market from 100 eggs, which is discouraging, but those twenty-five chicks are repeated the next hatch, and in twelve weeks we have sent two hatches to market, have twenty-five chicks ready to go in three weeks more, and another lot of twenty-five just out, after estimating for loss. In twelve weeks then we have 100 chicks, worth anywhere from $50 to $100, or more than the entire cost of the incubator and brooders. Bear in mind we gave the hens big hatches and made low estimates for the incubator.

ARE THE CHICKS EASILY RAISED?

In our liberality towards the hen we did not mention all the obstacles in her way, especially in winter. There are always a few strong chicks that keep her moving and one by one the weaker little fellows perish because she is too restless and they cannot nestle under her for warmth. Hawks, lice, rats, dampness and other troubles attend her, and if she raises, in winter, one-half of her brood, she will be lucky. The brooder makes every chick an independent individual. He can warm himself at will. The cat, rat, hawk, owl, and bad weather are defied. He "holds the fort;" can "laugh and grow fat," receiving all the care that can be bestowed, his owner knowing always where he is and what he desires. Not being afflicted with filth, vermin, and a scanty supply of food, he grows rapidly and is superior to the chick hatched under a hen from the time of leaving the egg till he graces the dish of some fastidious epicure. We know of a case in which 125 chicks out of 126 were raised in a brooder, but such good results are rare. It is best to estimate fairly, and make all due allowances. *Everything* depends upon the one who manages, and *that* is the secret of success.

COST OF RAISING.

There may be a little variation, but 100 chicks, when hatched, will weigh (together) five pounds, and they double in weight every ten days. Thus, at the end of ten days, one hundred chicks weighed together, should weigh ten pounds; at the end of twenty days, twenty pounds; thirty days should make forty pounds, and at the end of forty days, eighty pounds. After forty days they do not double, but grow rapidly, and sometimes weigh two pounds each when ten weeks old. The 100 chicks will eat as many quarts of cracked corn, or its equivalent per day, as they are weeks old. Or, in other words, they will consume one quart daily the first week, two quarts the second, three quarts the third, and so on to the tenth week, when they will have reached ten quarts. Thus, we have fifty-five quarts of feed for 100 chicks ten weeks, or about a bushel, which we may value at $1.00. The cost of a chick, then, is one cent a week for ten weeks, or ten cents. As the earlier ones are sold before they reach the age of ten weeks, the cost is less. The above is the test of *actual experiments* made, and is neither theory nor guesswork.

WHERE THE PROFIT COMES IN.

We have shown that it costs but 10 cents to keep a chicken ten weeks. Suppose it only sells for 25 cents, or about 16 cents per pound when weighing only a pound and a half, (and our estimate for feed was rather high, while our weight for ten weeks is low,) it does not require figures to show that even when the small sum of 25 cents is derived, that there is still 15 cents profit on every chick. We have, in order to compare, supposed that we put 100 eggs in the incubator and hatched only forty chicks; then we allowed a heavy loss of fifteen chicks out of the forty, leaving only twenty-five. We charged 10 cents each for raising the twenty-five, and allowed only 15 cents profit on each chick, and in twelve weeks we clear $15 on 100 chicks, but we are sure every one who reads this will acknowledge that twenty-five chicks raised from 100 eggs is a very low estimate, and no chicks sell so low as the price named except when they are hatched in midsummer, and then, of course, eggs are cheaper and the expense less.

WHY INCUBATORS SOMETIMES FAIL.

For the same reasons that hens fail. Eggs from fowls that are lacking in vigor will not hatch under hens nor in the incubator. Eggs from immature pullets, over fat hens, or from yards in which the cockerel is too young or the stock inbred, will not hatch. Eggs that have not been collected quickly in severe cold weather, and become chilled, will not hatch. Imperfections of eggs in size, shape and shell causes failure. the chief reason why incubators sometimes fail is that the *purchaser always knows more than the maker*, and will not do as directed. Incubators will hatch as many eggs (or more) as hens, and the chicks will be stronger, but they must be *managed for profit*, and profit is sure to result.

WHAT A SMALL AMOUNT OF MONEY WILL DO.

If $50 be invested in an incubator and brooders, in three months, on the low estimate we made of 15 cents per chick, the profit will be about 33 per cent on capital, deducting all expenses for feed; but if the chicks are hatched early, the capital will be returned in six weeks. It may be stated that in the warm season when prices are low, that eggs hatch better, more chicks will be secured, and less loss occur. In twelve weeks the original capital will be returned and a profit besides. Can anything be more favorable for a person with limited means? No investment can be better, and the occupation is one that will not only be pleasant and profitable, but be engaged in by quite a number who cannot find employment in other directions. Of course, however, to engage in poultry raising one should have suitable quarters and buildings, and enter into the business with the resolution and steady purpose that are necessary for any business. Do not put in the eggs and go to sleep, trusting to the incubator to do what brains should direct, but see that it does its work properly. An engine has a safety valve and a governor, but the engineer must be at his post also, and as the hatching of chicks is an infringement on nature, how much more should we be willing to do our whole duty in order to obtain good results therefrom.

TABLE OF PROFIT AND LOSS—WHAT AN INCUBATOR WILL PAY IN THREE MONTHS.

EXPENSES.

Cost of Incubator	$ 25.00
Cost 100 eggs, 3 cents each, four trials	12.00
Oil	3.00
Feed for 100 chicks	10.00
Four Brooders at $10 each	40.00
Total Expenses	$ 90.00

RECEIPTS.

One hundred chicks at 7 cents each	$ 75.00
Cost of feed, oil, and eggs	25 00
Profit	$ 50.00

The incubators and brooders, being permanent investments, their cost should be proportioned among all succeeding hatches. Suppose this method of hatching was kept up and more incubators added, a very large yearly income could be made. Many breeders are doing and others can do likewise.

HOW TO MANAGE AN INCUBATOR.

Having received the machine, unpack it carefully, read and study the directions, and if in doubt about any point, write to the manufacturers. Do not be in too much haste about putting the eggs in. Wait till you learn the working of the machine fully. If you have a good, dry cellar, that is the place to put the machine; if not, put it where the sun or wind will not strike it. The best results will be had by not burning a fire in the room. Having fixed the regulator, fill and trim your lamp. Now, fill the machine with water, till within one inch of the top; this gives room for expansion. Now, lay the thermometer on the egg-tray, about the fourth row from the back of the machine, and let me say right here, study the thermometer, so that at a glance you can tell what degree it is. Then fill the moisture pan or pans with warm water, and set them in their places.

The egg must have moisture, remember that, and if the machine does not supply it, you must. The reason so many fully matured chicks die in the shell is for want of sufficient moisture; the membrane interving between the chick and the shell will become so tough that the little orphan cannot break through it, and so dies. In putting the eggs in the drawer, put the large end up. After putting the eggs in the machine, do not change them for three or four days.

Attend carefully to the lamp. The wick does not need trimming every night; rub off the burnt wick with a stick or with a knife. The wick will last three times as long by doing this. Keep the burner clean; an incubator lamp is so near the floor that it catches a great deal of dust.

A hen after sitting a week or ten days, loses much of her natural heat, from causes not necessary to mention here; but at the same time, changes are taking place in the egg. Circulation of the blood has commenced, so there is generated in the egg a vital heat that balances the loss of temperature sustained by the hen. So taking this as a guide we must reduce the temperature of the egg-chamber as the heat in the egg increases. The nearer the time for hatching the more moisture the eggs need. When the chicks begin to pick the shell, keep the machine closed, and do not disturb them at this time. Nearly every one wants to pull out the drawer and watch the little fellows work out; but don't do it, or you will chill them so they will never get out.

HOW TO MAKE AN INCUBATOR.

Here is the rule for making an incubator that I took from a poultry paper and have found very successful. Experiments with the incubator here given have been made all over the country. It is one that is in actual use and has always given satisfaction.

To make this incubator, get your tinner to make you a tank fifteen inches wide, thirty inches long and twelve inches deep, of galvanized iron or zinc, the iron being preferable. On the top should be a tube one inch

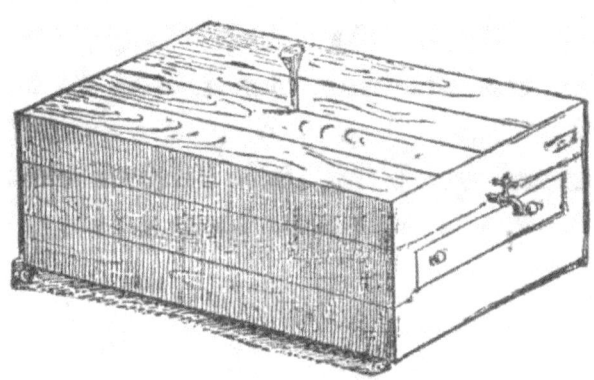

Fig. 1. THE HOT WATER INCUBATOR.

in diameter and eight inches high. In front should be another tube, nine inches long, to which should be attached a spigot, as illustrated in 2. Having made your tank, have what is called the ventilator made, which is a box with a bottom but no top. The ventilator should be eight inches deep and one inch smaller all round than the tank, as the tank must rest on inch boards. In the ventilator should be four or six tin tubes, one-half inch in diameter and six inches long. They should extend through the bottom, so as to admit air from below, and to within two inches of the top, or a little less. Now make an egg drawer, which is a frame of wood, three inches deep, having no top or bottom, except that the front should be boxed off and filled with sawdust, which is covered afterward with a piece of muslin, to keep the sawdust from spilling. This box in front of the drawer exactly fits the opening in Fig. 3, when the egg drawer is in its place. Of course the egg drawer must be made longer than the tank and ventilator, in order to allow for this space which it fills in the opening, which is the packing all around the incubator. The bottom of the egg drawer should be made by nailing a few slats lengthwise to the under side, or rather, fitting them in nicely, and over the slats in the inside of the drawer, a piece of thick, strong muslin should be tightly drawn. On this muslin the eggs are placed in the same position as if laid in a hen's nest, and it allows the air to pass through to the eggs for ventilation. The eggs can be turned by hand, marked for designation, or an egg turner may be made by fasten-

Fig. 2. THE TANK.

Observe that the tubes on the top and spigot are quite long, in order that they may extend through the packing of sawdust which is to surround it. This tank is to have a close-fitting covering (top and sides) of wood, to resist pressure of water. The bottom is not to be covered.

ing slats crosswise to one on each side running lengthwise, something like a window lattice, and when the eggs are placed between these slats, by merely pushing the frame the eggs will turn over exactly on the same principle that an egg will roll when it is pushed by a block, a book, or anything else, but we believe the method is patented, and do not advise infringement.

Having prepared the tank, let it be covered with a box, but the box must not have any bottom. This is to protect the tank against pressure of water on the sides, and to assist in retaining heat. Such being done, place your ventilator first, egg drawer next, and tank last. Now place a support under the tank and the box, or have them rest on rods, and as the weight of water will be great in the center, the iron rods should be placed crosswise under the tank every six inches. Now fasten the three apartments (ventilator, egg drawer and tank) together, with boards nailed to the sides and back and front, (of course leaving the opening for the egg drawer), care being taken to drive no nails in the egg drawer, as it must move in and out, and should have a strong strip to rest on for that purpose. Having completed these preparations, make a larger box to go over all three, so that there will be a space on the sides, back, front, and on the top, but as the ventilator must be filled with sawdust to within one inch of the top of the tubes, it serves for the bottom packing. Make the outer box so that there will be room for filling all around the inside box with sawdust, and also on the top, being careful to let the tube for pouring in the water come through, as also the spigot in front. About four inches or so thickness of sawdust is sufficient, according to preference. The front of the incubator must be packed also, but an idea of how it should be done may be learned by observing the opening in Fig. 3, which is so constructed that the box in front of the egg drawer, 4, exactly fits into it, and completes the packing when the drawer is shut. The incubator should be raised from the floor about an inch, when completed, to allow the air to pass under and thence into the ventilator tubes.

The incubator being complete, the tank is filled with

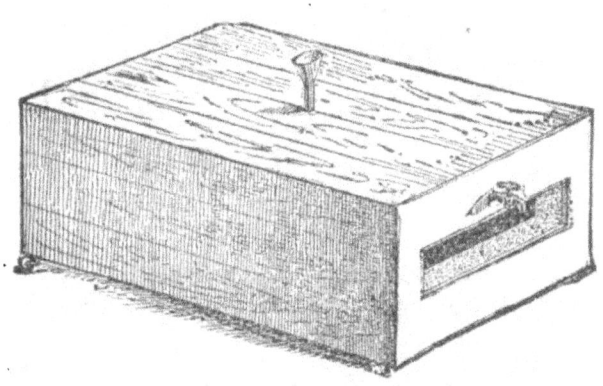

Fig. 3 DRAWER OPENING.

Shows the thick packing, which is noticed at the opening. This extends all around. The front of the egg drawer (Fig. 4) fits in its place in order to complete the surrounding packing, when the incubator is closed, as at Fig. 1.

boiling water. It must remain untouched for twenty-four hours, as it requires time during which to heat completely through. As it will heat slowly it will also cool slowly. Let it cool down to 120 degrees, and then put in the eggs, or, what is better, run it without eggs for a day or two in order to learn it, and notice its variation. When the eggs are put in, the drawer will cool down some. All that is required then is to add about a bucket or so of hot water once or twice a day, but be careful about endeavoring to get up heat suddenly, as the heat does not rise for five hours after the additional bucket of water is added. The tank radiates the heat down on the eggs, there being nothing between the iron bottom of the tank and the eggs, for the wood over and around the tank does not extend across the *bottom* of the tank. The cool air comes from below in the ventilator pipes, passing through the muslin bottom of the egg drawer, to the eggs. The 15x30

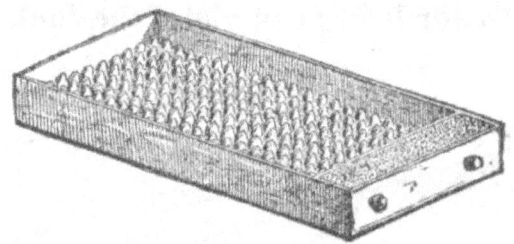

THE EGG DRAWER.

The space just in front of the egg is the portion partitioned off to fit in the opening at Fig. 3. Th egg drawer is therefore longer than the tank and ventilator.

inch tank incubator holds 100 eggs if turned by hand, but less if the eggs are placed between slats. Lay the eggs in the same as in a nest, promiscuously.

DIRECTIONS.

Keep the heat inside the egg drawer as near 103 degrees as possible; the third week at 104 degrees. Avoid opening the egg drawer frequently, as it allows too much escape of heat. *Be sure your thermometer records correctly*, as half the failures are due to incorrect thermometers, and not one in twenty is correct. Place the bulb of the thermometer even with the top of the eggs, that is, when the thermometer is lying down in the drawer. The upper end should be slightly raised so as to allow the mercury to rise, but the bulb and eggs should be of the same heat, as the figures record the heat in the bulb and not in the tube. Keep a pie pan filled with water in the ventilator for moisture, and keep two or three moist sponges in the egg drawer, displacing a few eggs for the purpose. Turn the eggs half way round twice a day at regular intervals. Let the eggs cool down for fifteen minutes once every day, but do not let them cool lower than seventy degrees.

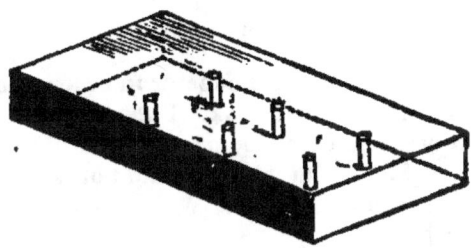

Fig. 5. THE VENTILATOR.

The tubes admit air from below, which passes into the egg drawer above through the muslin bottom of the egg drawer, to the eggs. The eggs rest upon the muslin, which is tightly drawn over narrow slats running lengthwise the bottom of the drawer.

No sprinkling is required if the sponges are kept moist. If the heat gets up to 110, or as low as 60 degrees for a little while, it is not necessarily fatal. Too much heat is more prevalent than too little. A week's practice in operating the incubator will surprise one how simple the work is. The tank will be troublesome to fill at first, but the matter will be easy after it is done, as it can be kept hot. Heat the water in two or more boilers, as a large quantity will be required, and pour it in through the tube on top of the incubator boiling hot, using a funnel in the tube for the purpose.

Just at the time of hatching out do not be tempted to frequently open the drawer. Cold draughts are fatal. Patience must be exercised.

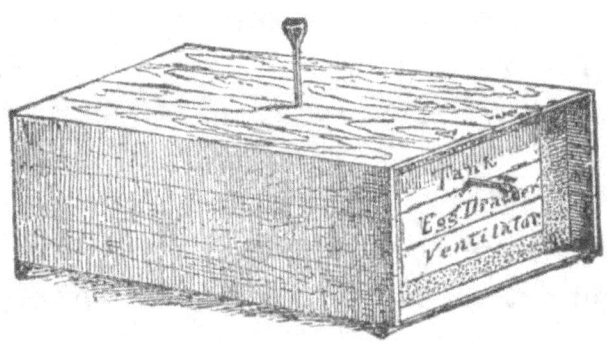

Fig. 6.

Here we remove the front of the incubator in order to show the positions of the ventilator, egg drawer and tank. First the ventilator, then the egg drawer (which of course should be longer than the others in order to fit in the opening shown at Fig. 3, but which we did not do here in order to mark the places), and on the top is the tank. When the front is completed the incubator is seen at Fig. 1.

DIRECTIONS FOR MAKING JACQUES' INCUBATOR.

In order to make the incubator herein described, you need only the following articles: A sugar barrel, a round tin clothes boiler about twelve inches deep, see Fig. 1, a tin milk pan, see Fig. 2, and a kerosene lamp

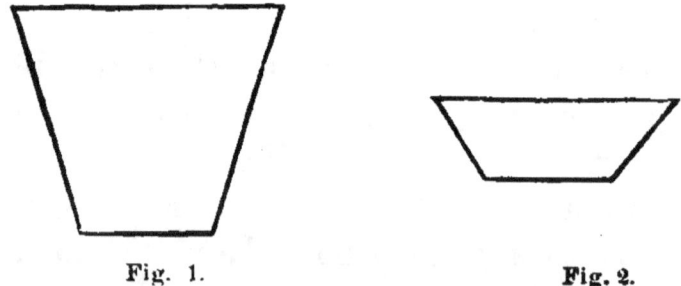

Fig. 1.   Fig. 2.

with a chimney. Have a barrel without a head, place in the boiler, which must be the size of the barrel so it

can be supported in its place by its rim resting on the chime of the barrel. The pan must be of such a size as when it is placed in the boiler (as in Fig. 3) it will have a space of about five inches between it and the bottom of the boiler. It will be necessary to solder the pan in this position. All the space between the pan and the boiler must be filled with water; this can be done by punching a small hole in the side of the pan near the top, and inserting a funnel. It will not be necessary to refill in three weeks, as the evaporation is so slow, you will not lose a quart. Make a door in the side of the barrel near the bottom, of sufficient size to admit the placing of the lamp under the boiler. Cover the outside of the barrel with four or five thickness of paper, well pasted on, to secure heat in the barrel. Bore two one-inch holes in the lower part of the barrel, one on each side, with tubes running from them to the base of the burner to the lamp, in order that the lamp may have a supply of oxygen to support the flame. Bore three one-inch holes near the top of the barrel, to allow the gas to escape. The cover must be lined and wadded, so it will fit tight to the boiler, that the heat cannot escape. Cut a hole in the cover 3x4 inches, paste a piece of glass over it; directly under place a thermometer, which can lie on the upper shelf of eggs, with the bulb touching a fertile egg; then you can ascertain the temperature without removing the cover; also bore two one-inch holes through the cover, insert a tin tube in each for the purpose of ventilating the egg chamber, which is of sufficient depth to allow three layers of eggs. Cover the bottom pan with a thin layer of cotton, on which the first layer of eggs, and at equal distances around the edge of the pan put three blocks of wood about two inches square, on which place a round sieve with one-half or three-quarter inch meshes; on top of this put another sieve larger than the first, so the rim of the lower one will support it. Cover the bottom of the sieves with pieces of coffee bag

or some other light material, so the heat can pass through it. The tubes to supply the lamp with air can be made by wrapping a piece of hardware paper around a broom handle three times, and pasting it together; after the paste becomes dry, slip off. Use common flour paste, well boiled.

DIRECTIONS FOR USING THE INCUBATOR.

Place the lamp under the boiler, turn on a good flame, when the mercury runs up to 100, reduce the flame so it can just be seen above the cone of the lamp, keep the temperature 103 the first week, 100 the second and 98 the third week. It is very easy to regulate, provided the temperature of the room is not subject to much variation. In case of very cold weather, close the ventilators at night and place a heavy woolen cover over the whole incubator. As the eggs need a certain amount of moisture, they should be lightly sprinkled with warm water every day, or as often as is needed. A good way to ascertain the amount of heat in the egg chamber is by keeping in a small piece of skin of a salt codfish; this should never be allowed to get so dry as to crack when bending it, nor to be so moist as to become wet, but should always be so that you can easily bend it. After the fifth day examine all the eggs by holding them up to a strong light. If any are perfectly clear, remove them, as they are not fertile, yet they are just as good for culinary purposes. Do not place the eggs in until you have secured and are able to to keep the the right temperature. I use a large bracket lamp and and have to fill it only once in 24 hours. After the sixth day turn the eggs daily, which can be done by removing the sieves; this will give them an opportunity to cool, as in the case of a sitting hen off her nest. It seems to be decided that the best temperature at which to keep the thermometer is 105 degrees the first week, 104 degrees the second week, and 102 degrees the third week.

A person would need be an excellent carpenter to be able to make an incubator, and such can easily suit themselves as to the pattern designed, and purchase the dimensions of the originator. But the cheapest way to save bruised thumbs and swear-words is to buy the "machine" ready made, and the directions for running it along with it.

YOUNG CHICKS.

HOW TO MANAGE THEM WHEN IN THE BROODERS.

Young chicks, when just from the shell, are well fed by nature, as they absorb the yolk a few minutes before coming out. They therefore need no feed for twenty-four hours, and even thirty-six hours. It is best not to feed them until they are about thirty-six hours old, as the rest and warmth will by that time have given them strength and activity. The first meal should be the yolks of hard boiled eggs, and the eggs should not be too stale. Lay the white of the eggs aside and feed to them for a change the next day. Be sure that the pieces are fine, in order that the chicks may easily pick them up. When the chicks are three days old, change the food to bread crumbs, moistened with milk, and allow coarse oat meal also. Continue this feed until they are a week old, giving a little chopped lettuce or cabbage once a day. If bread crumbs are not convenient, which is often the case when there are large numbers of chicks, make a thick porridge of oat meal and rice mixed, cool well, and let it become cold. Now beat up an egg in milk or buttermilk, or even hot water, and add to the porridge. Thicken the porridge with corn meal and feed to them. It must be borne in mind that no food is good for young chicks if continued, and hence it should be varied. After they are a week old, they should be given screenings, coarse oat meal, cracked corn, millet seed, sorghum seed, broken rice, or any other hard food that they can eat, but the soft food should also be given at least twice a day. Be cautious about feeding too much hard boiled egg. Give them plenty of clean water to drink. A good food after the second day is cold boiled rice, thickened stiffly with very fine oat meal, fine bran (ship stuff) and corn meal, equal parts. After mixing, warm it a little and feed. By all means get a bone mill or procure bone meal, and mix a little together with powdered charcoal, in the soft food. The common ammoniated bone meal will not answer. Button filings or bones from the table are preferable. Twice or three times a week, if convenient, after the first week, feed finely chopped meat of some kind, avoiding salt or pickled meats. Feed it cooked or raw. When green stuff is not procurable, a few onions (with tops,) lettuce, cabbage or young rye, chopped fine, will be relished. One of the best foods in cold weather, when green stuff cannot be had, is to take clover hay and cut it very fine (a tobacco cutter answers well for cutting it) and pour boiling water plentifully over the chopped clover and let it stand all night. The next morning boil it, adding potatoes, carrots, turnips, beets or anything you prefer. A small piece of beef, beef liver or ground meat may be put in. When well cooked, thicken with one part fine bran, one part ground oats and two parts corn meal, salting to taste. Now add a teaspoonful of fine bone meal and the same quantity of charcoal, and you have a cheap, nutritious, variable food, which contains all the elements necessary for growth, heat and health. Once a week parch some wheat, screenings, ground oats, or even corn meal, and feed to them, the meal being moistened. Put in some fine sand and gravel. The chicks must have something to scratch.

BROODERS AND THEIR MANAGEMENT.

It is not necessary for me to attempt to state which is the better plan for managing a brooder, as that is not within my ability. I may give my own ideas, but the experience of others may be different. I will, however, endeavor to throw as much light on the subject as possible.

I will give first directions for making a brooder, after Mr. Dakin's brooder as given by himself, he says: "The brooder I now try to explain will easily accommodate fifty chicks, and is two feet wide, four feet long and sixteen inches deep. Make a box this size without a cover, cut an opening six inches high and eighteen inches long (for the hot water can) in one side near the (a) end and close to the bottom. Put a partition in the box across the center (b), first cutting holes in the partition for the chicks to run through, the bottoms of the holes to be nine inches from the top edge of the partition, and the size may be three and one-half inches wide and five inches high. Above these holes I cut out for a strip of glass (c) two inches wide and eighteen inches long, for light in the "mother" in case the holes are closed. Fasten in your partition, which divides the box into two parts, each two feet square, lay a floor in the part opposite the one you intend for the water (d) can, and have the floor even with or a little below the bottoms of the holes in the partition. This completes the "run" part. The "mother" comes next. Make a box twenty inches wide, twenty-two inches long, nine inches deep, outside measure (e); this needs but one side and the two ends, and no top. Place this in the half of the large box back of the "run," with the open side to the partition, and have the floor of the "mother on a level with the floor of the "run." The "mother" may be held in position by placing posts of the right length, under it in the corners, and the top should be an inch below the top of the large box. And there should be an inch space between the two boxes on the two ends and back. Make hinged covers for the "mother" and fasten them to the top edge of large box. You will also need a wire screen over the run to prevent the chicks from flying out; and a frame fitted in the "mother" with thin muslin or cheese cloth tacked on loose enough to "bag" down, for the chicks to hover under, and a few small holes bored in the bottom of the large box, under the can, for the fresh air to pass through. Fit and hinge on a little door for the can opening first spoken of, and if you wish the brooder raised up from the floor you must put legs under it. Your brooder is now complete, with the exception of the can, which should be four inches deep and sixteen inches square, with an oil can screw top soldered in one corner. This can costs about one dollar. We heat the water by placing the can on the stove, first unscrewing the top or cap; otherwise steam will generate and burst the can. The principle of the brooder is that the heat warms the floor of the "mother," and also passes up the spaces between the two boxes and over the "mother," thus giving an even temperature throughout the same, the surplus heat passing out the entry holes into the run. On very cold nights we lay a newspaper over the run to prevent the heat from escaping too fast, but openings must be left for ventilation. In these brooders I am never troubled with chicks "piling up," and it is for

Fig. 1. DAKIN'S BROODER.

this reason that I favor under and over heat. If the floor is warm the chicks will spread out; if it is cold and the heat comes from above, they will climb up on each other to reach it, and of course the weakest chicks are at the bottom of the pile, and will require the "sad service" in the morning. I have tried to give the plan

of a cheap and good brooder that can be made at home by any one who can use a saw, hammer and plane. The cost, including the can, need not exceed three to four dollars, and will answer the purpose of any breeder who wishes to try raising a few chicks without the hen to care for them. The brooder may be increased in size and may be improved and made expensive. I have one with glass covered run and tight roof over the "mother" to shed rain, and heated by circulation of hot water forced through a coil of pipes by a lamp boiler. This brooder is mounted on wheels and can be moved from one place to another very easily. I would advise anyone attempting to make a larger size than the one I describe, to arrange the water can to be filled without being removed, and to have some non-conductor packed around the can to save heat.

HOW TO PRESERVE EGGS.

Next to good winter laying hens for profit, comes a good method of preserving eggs. There are a great many methods suggested for this purpose, and we will give not only the plan we pursue, but the experience of others. The best of all preservatives is sulphur, but as sulphur will not dissolve in water, we convert it into a gas by combining it with oxygen, forming what is known as sulphurous acid gas (not sulphuric,) which is done by simply burning it. We will say this much

in the beginning, however, which is that eggs so packed in a box as to permit them to be turned over daily will keep twice as long as those not so treated. By packing them in a box with oats as a filling, and then turning the box upside down, a large number can be turned at once. Another point is that eggs from hens that are confined in yards without the companionship of cocks keep better than under the reverse condition, or rather, infertile eggs keep better than those that are fertile.

THE SULPHUR PROCESS.

Take a common starch box with a sliding lid. Put the eggs in the box, and upon an oyster shell or other suitable substance, place a teaspoonful of sulphur. Set fire to the sulphur, and when the fumes begin to rise briskly, shut up the lid, making the box tight, and do not disturb it for half an hour. Now take out the eggs, pack in oats, and the job is done. If the oats or packing material be subjected to the same process, it will be all the better. If a barrel full is to be preserved, place the eggs in a tight barrel two-thirds full, with no packing whatever. Fire a pound of sulphur upon a suitable substance, on top of the eggs in the vacant space over them, shut up tightly, let stand an hour, and then take out the eggs. As the gas is much heavier than the air, it will sink to the bottom, or, rather, fill up the barrel with the fumes. In another barrel or box place some oats, and treat in the same way. Now pack the eggs in the oats, head up the barrel, turn the barrel every day to prevent falling of the yolks, using each end alternately, and they will keep a year; or, according to the efficiency of the operation, a shorter, or even a longer time.

THE LIME PROCESS.

Slack one bushel of stone lime and add water to make sixty gallons, to which add four quarts of salt. Stir at intervals and allow to settle. Dip off the perfectly clear liquid and pour in a cask to the depth of about fifteen inches; place eggs in liquid to about one foot in depth, and add some of the pickle that is a little milky. Add more eggs, over which pour more of the milky pickle. When the eggs are within about four inches of the top of the cask, cover with a piece of cotton cloth, and upon the cloth spread two or three inches of the lime that settled after the slaking. It is important that there always be sufficient pickle to cover the lime on the cloth. The object of placing the lime on top is to keep the water constantly saturated with it. Should some of the dissolved lime be changed into carbonate, more is at hand to be dissolved. We are not aware that the chemistry of the process, including the action of the lime water upon the egg shells, which are themselves carbonate of lime, has been studied. The subject is one worthy of careful investigation. For placing the eggs in the pickle, a tin basin or an old stew pan, punched full of holes, and the edges covered with leather is necessary. Fill with eggs, immerse in the pickle, then turn out slowly; the eggs will fall out and settle gently without breaking.

When limed eggs are sent to market, they are washed, dried, and packed in cut straw.

This process can be used without the salt if preferred.

THE BEESWAX PROCESS.

A French authority gives the following method: Melt four ounces of clear beeswax in a porcelain dish over a gentle fire, and stir in eight ounces of Olive Oil. Let the resulting solution of wax in oil cool somewhat, then dip the eggs one by one into it, so as to coat every part of the shell. A momentary dip is sufficient, all excess of the moisture being wiped off with a cotton cloth. The oil is absorbed in the shell, the wax hermetically closing all the pores. It is claimed that eggs thus treated and packed away in powdered charcoal in a cool place, have been found as fresh and palatable as when newly laid.

PARAFFINE,

which melts to a thin liquid at a temperature below the boiling of water, and has the advantage of being odorless, tasteless, harmless and cheap, can be advantageously substituted for the wax and oil, and used in a similar manner. Thus coated and put into lime pickle, it is said that the eggs may be safely stored for many months; in charcoal, under favorable circumstances, for a year or more. They should be kept in a cool place.

THE SCIENTIFIC AMERICAN PROCESS.

Having filled a clean keg or barrel with fresh eggs, cover the eggs with cold salicylic water. The eggs must be kept down by a few small boards floating on the water, and the whole covered with cloth to keep out dust. If set in a cool place the eggs so packed will keep fresh for months, but they must be used as soon as taken out of the brine. To make the salicylic solution, dissolve salicylic acid (which costs about $3 a pound) in boiling water, one teaspoonful of acid to the gallon. It is not necessary to boil all the water, as the acid will dissolve in a less quantity, and the rest may be added to the solution cold. The solution or brine should at no time come in contact with any metal. In a clean, airy cellar one brine is sufficient for three months or more, otherwise it should be renewed oftener. For that purpose the kegs, etc., should have a wooden spigot to draw off the liquid and replenish the vessel. Butter kneaded in the same solution and packed tight in clean stone jars will keep fresh the whole winter, but must be covered with muslin saturated in the water, renewing it sometimes. Cover the jars with blotting paper saturated with glycerine. Salicylic acid is harmless and yet one of the best and certainly most pleasant disinfectants in existence, with no color nor taste. The water is an excellent tooth wash and the best gargle to prevent diphtheritic contagion.

WAX PAPER PROCESS.

At the Birmingham, England, show, there was a competition for the best dozen of preserved eggs. The eggs had to be sent in to the custody of the secretary prior to October 1, so that at the time of examination by the judges they had to be at least two months old, and it was objected by many that the time was too short and ought to have covered six months. The plan of testing adopted by the judges was as follows: To break one of each set into a clean saucer, then to bring the best eight together in the same saucepan, putting them into the cold water and removing from the fire as soon as boiled, and allowing them to remain one minute and a half before tasting. Another set, one from each selected dozen, were boiled ten minutes and opened when cold. Those preserved in lime water were not satisfactory, milk of lime being more highly recommended. Others that had been coated with melted drippings or beeswax were also found wanting, the whites being thin and watery. The best had been simply packed in common salt. These had not lost sensibly by evaporation, had good consistent albumen, and tasted best when boiled. The eggs that took the second prize were adjudged nearly as good as the first. The young lady who packed them gave the following as her method:

Melt one part of white wax to two parts of spermaceti, boil and mix thoroughly; or two parts clarified suet to one of wax and two of spermaceti. Take new laid eggs, rub with antiseptic salt or fine rice starch. Wrap each egg in fine tissue paper, putting the broad end downwards, screw the paper tightly at the top, leaving an inch to hold it by. Dip each egg rapidly into the fat heated to 100 degrees. Withdraw and leave to cool. Pack broad end downwards in dry white sand or sawdust.

It was generally believed that had the contest covered a longer period these would have stood first. Another point of superiority in this last method was the fine appearance of the eggs, the shells being pure and clean as when first laid. For home use, probably the common salt method, owing to its simplicity, will be generally preferred, but for market, doubtless, the extra pains required by the second method would pay.

ANOTHER.

Use waxed paper, in which wrap each egg, as oranges are wrapped in tissue paper for shipping, pack small end down, fill in between with white sand, head up the box or keg, keep in a cool place, and turn the box over ocasionally.

THE HAVANA PROCESS.

Take twenty-four gallons of water, put in it twelve pounds of unslaked lime and four pounds salt. Stir it well several times a day, and then let it stand and settle until perfectly clear. Then draw off twenty gallons of the clear lime and salt water. By putting a spigot in the barrel about four inches from the bottom you can draw off the clear water and leave the settlings. Then take five ounces baking soda, five ounces cream of tartar, five ounces salt peter, five ounces borax and one ounce alum; pulverize these, mix and dissolve in a gallon of boiling water, which should be poured into your twenty gallons lime water. This will fill a whisky barrel about half full, and a barrel holds about 150 dozen eggs. Let the water stand one inch above the eggs. Cover with an old cloth, and put a bucket of the settlings over it. Do not let the cloth hang over the barrel. After being in the liquid thirty days the eggs may be taken out and packed in boxes and shipped. Do not use the same pickle but once. You need not wait to get a barrel full, but put in the eggs at any time. As the water evaporates, add more, as the eggs must always be covered with the liquid. It does not hurt the eggs to remain in the pickle. It is claimed that this process will keep them a year.

Let me append right here a sentence taken from a poultry book that I purchased early in my chicken experience:

"The Havana receipt for preserving eggs will be sent to any address upon receipt of $1; send $1 bill, Postal

Note, Money Order, Registered Letter or Draft." No comments are necessary.

THE POULTRY PROCESS.

To keep eggs the "year round," take one pint of salt and one quart of fresh lime, and slake with hot water, When slaked, add sufficient water to make four gallons. When well settled pour off the liquid gently into a stone jar. Then with a dish place the eggs in, tipping the dish after it fills with the liquid, so they will roll out without cracking the shell, for if the shell is cracked the egg will spoil. Put the eggs in whenever you have them fresh. Keep them covered in a cool place and they will keep fresh for one year.

DRY EARTH PROCESS.

A correspondent writes: "Last summer I was induced to try packing down eggs for the winter use. I had in seasons previous limed them, but a limed egg is not altogether to my taste. Last summer I took sweet, clean kegs, set them in a cool, dry place, with a barrel of powdered dried earth near at hand. In the kegs I placed a layer of this earth, then a layer of eggs, small end down, then a layer of earth, and so on until the kegs are filled. These eggs were quite good six months after packing down. By placing the eggs small end down the yolk is prevented from dropping down on the end and settling on the shell, while the dry, fine earth keeps them from the air. I suppose 'that ashes or bran or any other fine, dry substance is as good as the baked earth, but I write only of what I have actually experimented with."

LINSEED OIL.

Smearing the shells with linseed oil is reported to be a good way. Rub the oil over the egg with the tip of the finger and suffer it to become dry on the shell. Eggs rubbed over with flax seed oil in three months lost four per cent., and in six months four and a half per cent. of their weight, and when opened were found to be fresh, with the smell of fresh eggs. Eggs not so treated lost eleven per cent. of their weight in three months, and in six months thirteen per cent.

PRESERVING WITH SALICYLIC ACID.

Coat the eggs with butter containing a few grains of salicylic acid; place in sawdust, the eggs not touching each other.

BORACIC ACID.

A mixture of boracic acid and borax dissolved in water is said to be excellent.

PRESERVING WITH WATER GLASS.

Coat the eggs with liquid silicate of soda (water glass) and pack in bran.

SHIPPING EGGS.

Always sell eggs as near home as possible. In the winter of 1884-85, eggs were shipped from Iowa to New York city, and there sold for just one-third the selling price in any city in Iowa. This is not a rare case, indeed it is so common that I offer the suggestion.

CAPONISING FOWLS.

HOLDING THE FOWL.

The object of caponising is to improve the quality and increase the quantity of the flesh of fowls. A capon will outgrow a cock of the same age, growing to nearly the size of a turkey, and being so quiet, their growth is produced with far less feed than other fowls, and as the flesh is extremely delicate and juicy, they command prices from thirty to sixty per cent. higher than the common bird.

All instruments used in the operation can be purchased and consist of a pair of crooked concave forceps, pointed hook, a pair of tweezers, and a steel splint with a broad, flat hook at each end.

Remove the feathers upon a spot a little larger than a watch, at a point upon the line between the thigh and shoulder. Next, pull the skin backward, so that it may slip forward again after the operation is completed, and with a keen knife make an incision an inch and a half long parallel with the last two ribs, and between them until the intestines are visible, taking care not to injure the latter. Now separate the ribs by attaching one of the hooks to each, and allowing the ends of the splint to spread, as they will do when let go. The intestines may be pushed away with an old teaspoon handle or other flat, smooth instrument, and when the testicles are found (attached to the back), the tissue which covers them must be held by the tweezers, and torn open with the pointed hook. Next grab one of the testicles with the crooked concave forceps, and with the tweezers lay hold of the spermatic cord, to which the testicle is attached. Now twist the testicle off with the crooked concave forceps, after which the operation is repeated on the other testicle; the incision must now be closed,—no sewing of the parts being required—the skin allowed to resume its place, and the feathers which were removed should be stuck on the outside and left to adhere by means of the blood, thus forming the only bandage necessary.

Take pains not to disturb the parts to which the testicles are attached. The pressure of the tweezers tends to prevent pain and loss of blood.

There need not be more than six or eight per cent. of the birds killed, even by a beginner, if he use care, and as these die by bleeding to death, they are just as fit for the table as when killed in the usual way.

In order to avoid bleeding, be careful not to rupture

the large blood vessels attached to the organs removed.

The best age for cockerels to be operated on is three or four months. In order that the intestines may not be distended, prepare the bird by shutting it up without food or drink for thirty-six hours previous to the time of performing the operation.

As capons continue to grow fat for a long time, they should be kept until twenty months old to gain the full advantage of the operation.

After the operation give the bird plenty of water, but feed very sparingly with soft, cooked food until they begin to move about freely and scratch.

Perhaps the better way for a beginner to do would be to practice a few times on dead fowls, then try on fowls that are to be killed, and when a little skill is attained, take cockerels intended for capons. The process must be thorough, or the results will not be satisfactory, as the bird will, after a little, grow quarrelsome and crow. In this case the fowl is called a slip. Slips command better prices than ordinary fowls, but not as good as capons.

BREEDS OF FOWLS.

Our pure bred fowls are, strictly speaking, not pure bred, but strains that, having been bred straight for many years, produce chicks like the parents, and knowing first just what we want in a fowl, we are able to procure such and such requisites in certain strains. Our majestic Brahmas and Cochins are, one and all of them, descendents of the old Cochin China, or Shanghai, imported from China about 1847. As a general thing it takes ten or twelve years to get a new variety to breed straight enough to entitle it to the name of breed.

Fowls are divided into three classes, respectively, as Thoroughbreds, Crosses, and Dunghills.

The first is one bred straight for a number of years. A Cross is a chick of a pure bred cock and hen, of two distinct varieties. And a Dunghill is a mixed breed of no particular kind or color.

In giving descriptions of breed, I have not entered into the minute details of the Standard, as fanciers will buy a copy of the Standard, and to other poultry raisers the minutia would be but verbiage and require more room that can be given in this work.

PLYMOUTH ROCKS.

The Plymouth Rock fowl is perhaps the favorite in America at the present time. As table fowls they are fine-grained, juicy and tender. As spring chickens they feather early, and mature quickly. As market fowls they are unexcelled, having plump bodies, full breasts, bright yellow skin and legs. They are good layers and can be depended upon for eggs all the year round. From what we can learn from the best authority, they originated from a cross between the American Dominique cock and the Black Java hen some thirty years ago. Being an American breed, they are hardy, and adapt themselves to any location. They are excellent mothers, being kind and gentle, and sufficiently light to avoid clumsiness. If it is found desirable to raise show birds for fairs or the like, I would suggest mating

light-colored cockerels with pullets of clear, dark plumage. It is needless to state the reasons for so doing, but if one wants satisfactory show birds, this rule must be strictly followed.

When first hatched the chicks are a soft mouse color, but when feathered are a handsome blue gray and white. Average weight at maturity: Cocks, 9 to 12 lbs.; hens, 7 to 9 lbs.

THE WYANDOTTES.

This breed originated in this country in recent years, and was admitted to the Standard in 1883. In color they are black and white, marked as in illustration.

They have a small rose comb, close fitting, face and ear lobes bright red. They are a very hardy, attractive and profitable fowl. Will lay as many eggs as any variety, with the exception of the Leghorns. Occasionally chicks will have single combs, but this is in their blood and dates back to their ancestors. As a breeder weeds them out, he gets less each season. These fowls are easily kept, and for the village, city or country, will prove themselves one of the very best of known breeds. Weight when full grown: cocks, $8\frac{1}{2}$ lbs.; hens, $6\frac{1}{2}$ lbs.

THE LANGSHANS.

Within the last few years the Langshans have come

to the front as desirable fowls. Being almost as large as Brahmas, they make good market fowls.

In color of plumage the Langshans are a rich bronze-green black; they resemble the black Cochins, but are a distinct breed. They are round and deep in body, with breast broad, full, and carried well forward. They attain maturity early and grow to a large size. Their skin is a pinky white, and they are an excellent table fowl, their meat being of a delicate flavor. They lay

better than any other Asiatics; are hardy, withstanding readily the most severe winter weather; they attain their growth quite as quick as any of the other Asiatics; they lay large eggs, are good winter layers, and are not stubborn sitters. They are everything that is claimed for them, and are the coming breed in the estimation of many breeders. Weight of full grown cock, 10 lbs.; hen, 8 lbs.

THE BRAHMAS.

The Brahmas possess many excellent points to be considered by all poultry raisers. They are large bodied, fine layers of good sized eggs, very tame, and therefore no wanderers from home. They are strong and healthy, enduring cold weather admirably. They are persistent sitters, but a trifle heavy and clumsy, and for that reason apt to break any of the thinner shelled eggs if allowed to sit on them. As they are very poor flyers, a fence two feet high will keep them in their grounds. They will thrive splendidly, if given dry quarters, and a fair amount of sunshine.

LIGHT BRAHMAS.

This breed is too well known to need any extended description; they are the largest of the Brahma family, and where meat is desired are good birds to breed; they make broilers early and are fair winter layers. The pullets will lay in the winter, if provided with good warm quarters. They are easily confined, a three foot fence around a small yard being sufficient, which makes them well suited for village or city. I can confidently recommend them to all who wish to raise poultry for market. Weight when full grown; cocks 12 to 13 lbs.; hens, 10 to 11 lbs.

DARK BRAHMAS.

In size the Dark Brahmas are second only to the Light Brahmas. They lay nearly every day in winter if given proper care, and if pure bred, scarcely ever sit

until they have laid thirty or forty eggs. They are in fact the champion winter layers. In color the hens are steel gray, each feather beautifully penciled similar to a Partridge Cochin, The cocks differ very materially from the hens in color. Their breasts being solid black or mottled with light neck and hackle feathers.

No breed is more hardy from the time chicks pick the shell, to ripe old age. This variety has not been known as long, nor disseminated as widely as the Light Brahmas, but in no respect is it the inferior, that we know of. Some fanciers think the Dark Brahmas superior for fattening and for capons. It is said they pluck nicer, fatten more evenly, and their skin shows a brighter yellow color when dressed. They bear confinement as well as do the Cochins, but being more active are not so liable to get out of condition from over feeding. The old way discovered by English fanciers of mating for breeding is correct. That is, black-breasted cocks, with rather dark colored hens for cockerels, and mottled breasted cocks with steel gray hens for pullets. Not understanding this principle of mating causes many breeders to fail, and give the breed up in disgust. They make, too, a valuable cross with a Creveceur or Dorking cock. To sum them up they are good foragers, layers and sitters. It is one fault with the breed that they will lay soon after being put out with their broods, and continue to deposit their eggs and mother their offspring at the same time. They are equally as good for winter laying as the Light variety. They mature a little earlier and their flesh for table use is generally preferred, as it is somewhat finer in texture. They have no superiors among the Asiatics and few equals among domestic fowl. Average weight: Cocks, 11 lbs.; hens, 9 lbs.

COCHINS.

The Cochins possess the valuable quality of being winter layers, and if of right age and properly managed, will lay eggs and raise chicks in mid-winter. The eggs are of the rich cream color, so much prized by consumers in one or two of our largest eastern cities, and are of good size. The Cochins are not restless if given but little space, and are very tame and gentle.

PATRIDGE COCHINS.

Among the Cochin varieties the Patridge stands the most prominent in our poultry exhibitions, disputing

with the Brahmas, Leghorns and Plymouth Rocks for the first rank. They are not only large, but very compactly built and heavily feathered, possessing fine carriage and elegant shape. The cocks are black on the breast, with full flowing hackles of a brilliant red color. The tail and fluffs are black; and the legs yellow. When in full plumage the blending of colors makes a handsome appearance, which is not easily surpassed. The hens are handsome, with reddish gold tinge on the neck, while the breast and body are of a rich brown color, distinctly and handsomely pencilled with a darker brown, and although the plumage of the cocks and hens is apparently very dissimilar, their uniformity in all other respects is easily noticeable.

The Partridge Cochins are exceedingly hardy, being able to stand not only severe winters, but the summers also; are of an extremely quiet and domestic nature. A very low fence three or four feet high prohibits their straying. They are very patient of confinement and therefore suited to people living in villages and suburbs of cities, when but little space can be allowed for poultry yards. They lay well and are good sitters, remaining close to the nest during incubation; and although they are accused of crushing the eggs and chicks, much of the difficulty arises from their being compelled to jump on the nests from above instead of passing in from the front. They are the best and most careful of mothers, and tenderly nestle and scratch for their broods until the chicks are quite advanced in age. When cocks of this breed are crossed on ordinary common hens, the produce is almost entirely like the sire. This is due to the breed being an old established one, and the fact that the chicks are always uniform in color and shape, shows the value of the Partridge Cochins as one of the best pure breeds for improving common flocks. They are not subject to roup and other ills brought on by wet and snow storms as some other birds. The chickens are hardy and can be raised without any difficulty. They are one of the best varieties for laying eggs in winter if properly managed. Pullets will often commence laying in the fall and lay all winter. With this variety you have good layers, heavy weights, fine plumage, symmetrical form, and in fact all the fine points which go to make an excellent fowl. Average weight: cocks, 11 lbs.; hens, 9 lbs.

BUFF COCHINS.

The Buffs are the oldest of the different varieties

of the Cochins. The beauty of these fowls consists largely in the uniformity of the rich buff coloring without markings, but the tendency of the plumage, which may be of a brilliant buff on young stock, to turn "mealy" in tint before they have reached a year, is an important feature to be met.

The standard weight for cocks is 10 lbs., and of hens 8 lbs.

BLACK COCHINS.

These fowls have many excellent qualities to commend them to admirers of the Asiatic breeds. They are hardy and considered by some the best winter layers among Cochins. They are not as large as the Partridge or the Buff Cochins, but are more easily bred true to feather, and like all black fowls, present a neat appearance.

WHITE COCHINS.

The White Cochin differs but little from the other members of the Cochin family, except in color, which is

in this breed pure white. They breed very true to color, but require a very clean house and smooth lawn in order to preserve neatness of appearance. The glare of the summer sun will also give their handsome white plumage a dull yellow appearance. Standard weight of cocks, 8 to 12 lbs; hens, 6 to 8 lbs.

THE LEGHORNS.

We know of no breed of fowls which can be better denominated as "egg machines" than the Leghorns, and farmers who live within a reasonable distance of our large cities are gradually finding out the fact that it pays to breed them for their eggs. As foragers they are unrivaled, and if they have their full liberty on a farm, will secure nearly their entire living during the summer, which tells of itself that they are destruction to gardens, and being good flyers, nothing short of a ten foot slat fence will hinder their progress. The cocks are regular gentlemanly highwaymen, with the greatest possible amount of braggadocio to the smallest amount of endurance in fight, their immense tender combs and wattles being easy points of attack to the enemy. They are also liable to frost bitten combs in severe weather.

WHITE LEGHORNS.

White Leghorns are the oldest of the Leghorn family, and are very beautiful fowls. They have pure white plumage, blood-red combs, yellow breasts, white ear-lobes, and orange-yellow legs. They are of medium size, beautiful in symmetry, and the carriage of the cock is very stately. While in disposition they are timid and suspicious, they can be easily tamed. Leghorns stand at the head of the different breeds for the number of eggs produced. Are non-setters and mature very early. Can be raised late in the season, after it is too late to start Plymouth Rocks, Brahmas or Cochins.

A Leghorn pullet often begins laying at four and a half months old, and cockerels often crow at six weeks old. They breed very true, and bear confinement well, although a free range is to be desired.

BROWN LEGHORNS

The Brown Leghorns are acknowledged to be the best layers in existence, laying if properly fed, a large egg. They are of medium size, and non-sitters. Are easily bred, being exceedingly hardy, feather very young, and mature early. My cockerels commenced crowing at six weeks old. They love freedom, and are rovers by nature, requiring very little feed, if any, where they have a free range. The pullets mature early, some often lay between four and five months old. The chicks feather out very young, looking like Bantams, so neat in feathering, lively and precocious. No one who has had any experience at all with Leghorns, will deny the little brown hen the palm over all others in egg production.

It is not from fancy alone that I would breed them, however, but from the fact that the Brown Leghorn fowl unites the beautiful and the useful to a greater degree than any other variety. Cocks weigh four to five pounds, hens three to four pounds.

BLACK LEGHORNS.

This variety of the Leghorn family is of comparatively recent origin, and is not at all common. They are rather larger than the other varieties, and are not bred with the rose comb. Their admirers claim that they are superior as layers to the Brown or White Leghorns, but this seems hardly possible.

DOMINIQUE LEGHORNS.

These are of still more recent origin, and are the smallest of the Leghorn family. They do not excel the Brown or White varieties as layers.

BLACK SPANISH.

The Spanish are among the old established varieties. and deserve all the honors accorded them. They are particularly noticeable, because of the white face, which is strikingly peculiar to this breed alone. They are good layers and non-sitters. The egg is proportionately large. with a white, smooth shell, and of delicate flavor, which make it a most desirable market variety.
. The pullets usually lay at six months old and will

continue through the winter; but the hens rarely begin laying before January, after which, however, they seldom stop more than a day or two.

As table fowls they do not present so fine an appearance as their forms when covered with their glossy, green black plumage would indicate, and are not so juicy and highly flavored as are those of many other breeds.

Weight of cocks, 8 lbs; hens, 6 lbs.

BLACK JAVAS.

The Black Java is a black fowl, with purplish azure reflections; and the cock is a glossy, velvety black; plump and square, back broad and body deep; comb single and deeply serrated, standing erect in both sexes; with well-proportioned wattles. Their disposition is very quiet. They are excellent sitters and good mothers, although not very broody. The eggs are medium size and white, but not so pure a white as the Spanish or Hamburgs. The chicks are of a bluish black, with

whitish down about the breast and under parts; they are sprightly, and grow well with ordinary attention; feather soon; when six months old the pullets are unsurpassed in beauty, and are very attractive. Like all black-feathered fowls, the dressed bird is white, but as a table fowl it is excellent, inclining to be always fat with ordinary feeding.

MOTTLED JAVAS.

In plumage the Mottled Javas resemble the Houdans. In size they approach the Plymouth Rocks. For table use they are excellent. As yet they have not been bred long enough to be always true to markings.

HOUDANS.

To my notion the Houdan is the handsomest barnyard fowl in existence. They are the most hardy, best and popular of any of the French varieties, they are the very best of layers of the largest of eggs. They mature early, and are one of the best table and market fowls. They are not termed "high flyers," and are contented almost anywhere, though in disposition are lively and sprightly. They are good foragers and are adapted for a free range, especially on a farm, where they are virtually non-sitters, and in order to breed them successfully, keep a few of the Asiatics, or heavier breeds to do the sitting and rear the chicks, although they generally sit very well after the second year. Their eggs are generally fertile, and the chicks seldom die, except by accident. Their thick crests and beards serve well as a protection from frost in winter Generally speaking, the Houdans are called the pet of the poultry yard.

As regards economic and useful qualities, the Houdan has but few peers. It lays nearly as many eggs as the Leghorn and far larger. In meaty qualities the Crevecœur and the Le Fleche only equal it, and no other variety excels it. They are hearty, vigorous, of exceedingly rapid growth as chicks, and not at all subject to diseases. Used as a cross on large fowl, the Houdan cock has no superior, and we forsee the day when it will be one of the most popular fowls for home and market consumption.

CREVECOEURS.

This breed is of French origin, and is the favorite fowl of that country. The fowl is, in form, very full and compact, and legs exceedingly short, especially in the hens. They are the most quiet and contented of fowls, laying very large eggs.

The comb is in the form of two well-developed horns, surmounted by a large black crest; the wattles are full, and like the comb, a very dark red. The plumage is black. The merits of this breed are, quick maturity, good table qualities, large eggs, and the fact that it bears confinement splendidly.

The Crevecœur is too delicate to bear the cold of Northern winters.

LE FLECHE.

Almost a French edition of the Black Spanish, which fowl it resembles very strongly, excepting in comb. It has two spike-shaped prongs, like horns standing straight up. These three breeds are very popular in France.

JERSEY BLUES.

Thirty years ago these fowls, which originated in the state whose name they bear, were considered a very valuable fowl. In color they are a light blue, with short wings and tail; legs generally black. They are not extra as a table fowl, and are not prolific egg producers.

SULTANS.

These fowls which rather resemble the Polish, were imported originally from Turkey, and are scarce even there. In general habits they are brisk and happy-tempered, and fair layers of large white eggs. They are non-sitters, and small eaters. Their plumage is white, with full-sized compact tuft on the head, bearded as in the Polish; good flowing tail and short well-feathered legs. They have the quaint little ways and habits of Bantams, and take petting with a confidence that makes them favorites with their owners.

Of course, this variety is not suitable for purposes of general utility, and can only be recommended to persons who want an ornamental fowl, not given to hatching, and that will do but little harm to any well-kept barn.

The cock's spurs are peculiarily liable to grow very long when the fowl gets old, and so much curved that the point enters the leg and causes much pain. This should be guarded against, and if necessary the spur shortened sufficiently to prevent such consequences.

The Polands like the Hamburghs are properly classed as fancy fowls. They are handsome, but tender, and cannot hold their own in promiscuous flocks.

Polish fowls are non-sitters. They have blue legs and small combs, which are nearly hidden under the crest of feathers. As chicks they are a little delicate, and are not as easily raised as some other breeds, if hatched early in the season, before warm weather has really come. May and June, and even later, are the months to have the young chicks come out. Then, with the mild dry weather, they thrive as well as can be wished for.

Polish are easily restrained, and are well suited to small places where only a few fowls can be kept. They are most excellent layers, laying a large egg, and are very ornamental for the grounds of a country residence.

While at large, they are very plump, with full breast, and their flesh is very tempting and toothsome.

WHITE CRESTED WHITE POLISH.

The White Crested White Polish are unquestionably one of the most ornamental fowls bred in this country at the present time. Their pure white plumage, their large and ornamental crests, and their graceful forms, put them in the front rank for all lovers of beautiful birds. They are peculiarly domestic, in which they are unlike many other fowls, fond of being petted, and manifest pleasure at being noticed. They are hardy and healthy, and free from the diseases so liable to attack poultry. They are prolific layers, non-sitters, and well adapted to a small yard. In fact a very desirable breed for the village or city. As a table fowl they are unsurpassed.

WHITE CRESTED BLACK POLISH.

The White Crested Black Polish is a very odd looking fowl, with a large white topknot which contrasts beautifully with the glossy black plumage. They are non-sitters, excellent layers, and a great favorite with those who keep them and know them best. This variety is the most popular of all the Polish breed.

While the handsome and attractive plumage and peculiar markings of the White Crested Black Polish make them objects of interests to breeders and fanciers alike, they have qualities, practical and profitable, which commend them to all who breed poultry, as they are most excellent layers of good-sized eggs, and under proper management will lay as well as almost any other breed we know of. Like the Leghorns and Hamburgs in some respects, they need warmth, care, and attention, during the prevalence of cold weather, and if they do not have this accorded, they cannot and will not lay eggs during the winter and early spring and —in fact not till the weather becomes warm and pleasant. In one respect, however, they are different from the Leghorns, for, not having those large combs which the Leghorns have in such an eminent degree, they do not suffer so badly during severely cold weather, if accidentally or carelessly exposed to its influence. The large topknots afford a considerable protection to the birds about the head, though this is, where hawks abound, sometimes objectionable.

One feature, and a desirable one in breeding this variety of Polish, is that almost as soon as the chicks are dry from the shell the fancier can tell whether the birds will have the proper proportion of white in the in the crests or not.

SILVER AND GOLDEN POLISH.

These varieties are quite as striking as either the White or White Crested Black, and possess all the other desirable qualities. Of course the beauty of each bird consists, as in all speckled fowls, in the regularity and uniformity of markings. The average weight of the Polish is, male 7 lbs., female 5 lbs.

HAMBURGS.

The Hamburgs are very uniform in size, and are quite small, being a little smaller than the Leghorns, but rivaling them in egg production. They are persistent layers, never offer to sit; have rose combs that are not easily frosted; and are good foragers, hunting and seeking their own food when given the range of the farm, but not at all partial to confinement, as their repeated attempts to get over the tallest fences give strong evidences of their love of freedom. In order to hatch them, the eggs must be placed under common hens, owing to the disinclination of the Hamburgs to sit, and if not fed regularly and systematically they will not thrive, because of their great laying powers, which entails a heavy drain on the fowl. In order to succeed with them they must not only be well fed but on a

variety of food, which should include meat and ground bone. The reason is, that such fowls as Leghorns and Hamburgs are compelled to perform greater service than most other breeds, and if not supplied with meat, bone and lime in some form, soon become addicted to the vice of feather pulling, which is the worst evil that could befall flock of fowls.

BLACK HAMBURGS.

The Black Hamburgs are non-sitters, lay medium-sized white eggs and are the rivals of the Leghorns. Their brilliant black plumage; elegant, broad, rose combs; and handsome carriage place them high on the list as ornamental fowls. They are active and industrious where they have free range, and cost very little to keep.

SPANGLED AND PENCILED HAMBURGS.

Of these varieties the Golden Spangled and Silver Spangled are the favorites, and next the Silver Penciled. The Black Hamburgs are, perhaps, the hardiest of the varieties of this breed, and though beautiful in plumage, cannot compete with the spangled varieties where fowls are valued for beauty alone. They are not as rapid in growth as might be desired or expected, but when fully matured will lay as many, and some breeders declare, more than any other breed, laying from 180 to 220 eggs in one year. The penciled varieties lay at about four months, the spangled at six months old. The eggs of the Hamburg are the purest white of any egg known, with pink sea-shell tinted lining, and though not large, are of fair size.

WHITE HAMBURGS.

In all points, except color of plumage, the White Hamburgs resemble the other varieties, and being one color, they can be bred more satisfactorily than the Spangled or Penciled varieties. Those who wish to raise fowls for their eggs, would do well to cross a White Hamburg cock, with light Brahma hens, the result being a cross breed fowl, which will make an excellent winter layer. All Hamburgs are a little sensitive to cold, and should be given good, tight houses. The chicks are best hatched a little late, as the damp and cold winds of spring are liable to give them roup. Until partly feathered they should not be allowed to run in damp grass in the early morning.

ANDALUSIANS.

These fowls are of Spanish origin, and have curious slaty-blue plumage. They are of fair size, and are said to be most excellent layers of large eggs which are of an exquisitely delicate flavor. The young chicks feather fast, and are quite hearty. Coming from a warm climate, this breed is not adapted to localities where the winters are long and severe.

This breed has not become very popular, although for no apparent reason as they possess some very valuable qualities.

The Andalusian must be considered a truly useful and handsome fowl; being, according to general testimony, the hardiest of all the Spanish breeds. This breed appears each year to increase the number of its admirers, and may, very probably, attain, in time, to a distinct class of its own.

BLACK RUSSIANS.

These fowls were introduced into the United States with considerable flourish of trumpets, some years ago, but they failed to sustain the reputation given them by those who imported them. They are decidedly rough-looking fowls, ornamented with tufts of feathers on each jaw and an abundant beard of feathers under the jaw. Iu Europe they are bred in white, buff and mixed colors, as well as black. They are fairly hardy, small eaters, and poor layers. In size they are about as large as the common barn-yard fowl. On the whole, we would not recommend the breed as one that will give satisfaction.

THE AMERICAN DOMINIQUE.

The American Dominique is a fowl that is sure to give satisfaction anywhere, especially so on the farm. They are very hardy, good winter layers, not easily frost bitten, having the rose comb, are of good size, plump and dress well, their yellow legs and skin recommending them as a market fowl.

DORKINGS.

To any one who has a moderate grass run and a dry locality, the Dorkings will give perfect satisfaction; but they are not suitable for confined spaces and damp yards. They are fair layers of large eggs; and for the table they are voted by epicures to be the best fowl in existence, excepting perhaps the Games. In size they are quite large, with full, plump breasts.

WHITE DORKINGS.

White Dorkings are an exceptionally beautiful variety, provided they have a clear grass range. The breed has of late years fallen into disrepute, from the fact that those who have bred them claim that they are delicate and

not good layers. Still, they have many admirers who deny these faults. They lay a pinkish egg, are almost non-sitters, of medium size, and as table fowls are unsurpassed. They have a rose comb, and are somewhat smaller than the other varieties.

SILVER GRAY DORKINGS.

This is undoubtedly but an offshoot from the colored variety and a fixed type of plumage made by careful selections. Colored Dorkings will occasionally throw silver gray chickens, and such are sometimes exhibited and bred with Silver Greys; but it is needless to say that disappointment is sure to ensue unless the strain has been kept pure for generations. They are as large as the Colored Dorkings and similar in all respects, except in the plumage.

It should be remembered that Silver Gray and the other varieties of Dorkings degenerate more than almost

any other fowls from interbreeding; and if fresh blood be not introduced they will rapidly decrease in size. They do not bear close confinement well, and they prove satisfactory only when allowed full liberty and where the soil and climate is dry.

THE GAME.

There is, to my notion, no fowl that lives any more ugly in appearance than the Game. I despise them, root and branch, but as I have attempted to describe different breeds, I find myself compelled to include these little reprobates, that delight in digging out each. others eyes, and then perching on the fence tops and boasting of it. But, as the old lady said when called on to take an oath, "If I must, I must, so here goes."

The recognized varieties of exhibition Game fowls are Black, Black-breasted Red, Blue, Brown-red, Silver and

Golden Duck-wing, Ginger Red, Irish Gray, Red and White Pyle, Spangled, and White. Of Pit Game fowls

BLACK-BREASTED RED GAME FOWLS.

there is an almost innumerable variety; among the best known of which are Claibornes, Kentucky Dominiques, Muff, Tassels, Strychnine, White Georgian, Shawl-necks, etc.

Breeding Pit Games—those for the pit exclusively—is a very fascinating occupation to many breeders; and the "champions" readily bring good prices. Pit Games are not, necessarily, of any particular breed, but are generally made-up from the staunchest, gamiest of other game breeds. Some breeders now have strains of pit games that cannot be excelled, but if any one wants to run a cock pit they can find literature enough on the subject, written by people who like Game fowls better than I do. For my part I should prefer to raise the wretched mop-rag looking chicks called

FRIZZLED FOWLS.

This is one of the most curious of fowls. The feathers all turn the wrong way, as if in rebellion against the laws of nature. This is most noticeable with the hackle feathers, but the entire plumage is more or less arranged this way. They are of no practical use, and are only bred as curiosities. They are of all colors and parti-colors.

RUMPLESS FOWLS.

Rumpless fowls are noticeable only for their entire want of tail, or any approach to one. There are Black, White and Speckled varieties, and they range in weight from 3 to 7 lbs.

SILKIES.

These are very pretty, interesting little creatures, but altogether fancy fowls, utterly useless for table or market. Their bones and flesh are coal-black, while, strange to say, their plumage is snowy-white, soft and silky, resembling spun glass. Their eggs, though small, are said to be excellent; they are pinkish-white in color, something like those of the Bantam.

The cock is a pattern of fidelity and gentleness; he assists his partner in the care of her family, and even acts as nurse. The hen is the best of foster-mothers for Hamburgs and Polish; their soft, warm feathering tempting these delicate little fastidious beings to look to them for protection.

The points of a Silkie cock are: plumage, white; crest

small, low, and set far back on the head; comb, dark-reddish purple; ear-lobes, turquoise blue; wattles, purple; legs, eyes, and beak, jet black; tail, full but not with many sickle feathers; body, low-set but large, and broad in the breast. Hen— precisely similar points, except the tail, which in her consists of a small bunch of feathers like marabouts. Her comb, also, is smaller, and crest, or tuft, larger.

BANTAMS.

And now, did I but wield the pen of a graceful writer, how gladly would I pay glowing tribute to my little pets. The White Bantam is my especial favorite, and if it is possible for so small a head to contain reasoning faculties, then one can understand how it comes that Bantams lack only the gift of speech to make them real people. However, lacking that, I suppose I must go on and describe them after the manner of descriptions of fowls generally.

At all poultry exhibitions, the pens of these little beauties are continually surrounded by admiring visitors. Not only from the ladies, whose eye for the beautiful is not to be questioned, but, on all hands are heard exclamations of praise; which, were they not so truly deserved, would certainly not be so willingly given. Their graceful form, beautiful plumage, sprightly bearing, and diminutive size have alone not made them such general favorites. Their contribution to the egg-basket, when compared with their demand upon the grain-bin, is certainly a good one. Their fine-grained flesh is delicious, while their contented disposition and limited amount of room they require, make them desirable, where more pretending, though perhaps less worthy varieties, could not be thought of.

No fowls will afford more pleasure to their keeper than these little pets. The clear, merry crow of the cock, his proud strut, and readiness to defend himself against all attacks, are a continual source of amusement; while the nimble little hen, ever on the lookout, and guarding her brood with the most unremitting care, cannot but be admired by anyone who will watch the graceful movement of her stylish little form. Of the many varieties kept at different times, none will give more pleasure and satisfaction. On account of their diminutive size, many at once class them as a delicate variety. This is, however, an error, which a little time given to the rearing of them will soon dispel.

BLACK BANTAMS.

In size are very diminutive; the plumage a rich black,

comb of the rose order, ear lobe, a pure white; are hardy and moderately good layers. The chicks are strong and healthy.

WHITE BANTAMS.

Except that the plumage and legs are white, this variety is identical with the black. They are very tame and can be taught all manner of tricks, and have wonderfully knowing ways. The hen will raise three or four broods of chickens in one season, if allowed her eggs.

PEKIN BANTAMS.

This pretty little Bantam, as might be supposed, is of Chinese origin. It is a diminutive Buff Cochin, but still shorter in the legs than the larger variety which it so much resembles. The chicks thrive and feather well and are generally full fledged when two months old, at which age they are very pretty.

In disposition, Pekin Bantams are quiet and gentle and much attached to each other.

SEABRIGHT BANTAMS.

The Seabright Bantam is of two varieties, called respectively Gold and Silver, from the ground color of the plumage; which, in one case, is a golden brown and in the other, a clear white. In perfect specimens every feather is laced or margined all around, as shown in the engraving. The tail of the cock has no sickle feathers, but is like that of the hen.

They are among the smallest and most beautiful of Bantams. The only objection to them is that their eggs are more prone than other varieties to be unfertile.

JAPANESE BANTAMS.

This interesting variety, which is said to have been brought from Japan, makes most pleasing pets for children. Their quaint ways, upright carriage, and the long streaming sickle feathers of the cocks, all combine to make them one of the most attractive kinds of Bantams.

They are very hardy, the hens good layers and attentive mothers; while the chicks mature early and are

easily reared. The only objection to them is, that they are apt to be too large if not hatched late in the season.

The plumage is white, with black tail, but the hackle has often a few black feathers in it. The comb is bright red, very large, and single, with deep serrations. The tail is very large, expanded, and carried so upright as, in the best specimens, to come in contact with the head. The legs are free from feathers, and bright yellow in color.

There is no Bantam that can be recommended more highly than the Japanese.

WHITE POLISH BANTAMS.

White Polish is a pretty little fowl, resembling in appearance the large variety by the same name, except, of course, in size, which is quite diminutive. They do no scratching to injure anything in a garden. They may be kept in the same yard with other fowls without any danger of crossing. In disposition they are quiet and gentle, and become, with proper treatment, very tame.

102 POULTRY RAISER'S GUIDE.

They are said to be good layers and excellent mothers.

We give a cut of the Yokohama fowl simply as a curiosity. The accompanying description was taken from THE POULTRY WORLD.

"It is the pleasure of the Japanese to produce, by artificial rearing, the abnormally long tails of the cocks, the growth of which is promoted during the moulting season by moist warmth, by burdening the feathers with stones, and the continual abode of the fowls on high bars. The cock has an astonishing long tail of about 20 feathers, ½ inch in breadth, and the longest measuring 13½ feet. This very remarkable kind of fowl, which has the longest tail of all bred in Japan, is not much known yet. It came originally from the province Iosa, in the Sihohu Island, and is also known by the name of Shinowaratou or Kuro sasa Oski. About sixty years ago the breeding of this remarkable kind of fowl was quite common in Iosa, and the breed has been much improved.

Various travelers who have visited Yokohama have been greatly struck with the immense length of the tail feathers of the cock birds found there and from time to time have written brief descriptions of the fowls."

GAME BANTAMS.

Are neither as pretty as the other varieties of Bantams and not as strong as the Games.

GEESE.

There are a good many kinds of geese; the only similarity between some of the species is that they are all two-legged. But the feathered variety is the only one in which there is any profit and of these we will rise to remark: Geese are by nature a water fowl but if forced to live on dry land they will do so, if given plenty of fresh drinking water. They begin laying when one year old, producing very large, white eggs, having thick shells. The period of incubation is thirty days and a goose can cover ten eggs easily, of which she will take the best of care, turning them each day, and bringing sufficient moisture on her feathers to insure hatching. The gander also takes great interest, and stands by, guarding the sitter, and woe betide the small boy who ventures too near the nest.

TOULOUSE GEESE.

These are the largest geese known. The plumage is pure white on the back, but shading down into grey on the lower part of the body. They have reached the enormous weight of sixty-eight pounds per pair. They are very hardy, are not noisy, and are easily raised. Their heavy bodies admit of their being confined by a low fence.

The geese lay thirty to forty eggs each in a season, and seldom offer to sit. We find them good to hatch, easy to raise and much stronger when young than common goslings. They grow so rapidly, that at four weeks they will weigh from six to eight pounds each; and at three months, fifteen to eighteen pounds. They yield half a pound of feathers to a "picking." They are small feeders for their size· and require no food but

TOULOUSE GEESE.

pasture, except in winter. In color, geese and ganders are exactly alike, viz.: a uniform, handsome gray, with breasts and under parts of body a shade lighter. They are gentle in disposition, not unruly, and can be fenced easier than sheep; breed at one year old; and in all respects are very profitable. The sexes can be distinguished by their forms and voices: ganders are taller, more upright, with larger necks, and gobble in higher, finer, and more rapid tones than the goose, the voice of which is low, deep bass and slow.

EMBDEN GEESE.

Embden geese are uniformly pure white and have prominent blue eyes. They are hardy and well adapted to any climate. The flesh, when cooked, is excellent. They attain large size, in some instances weighing fifty pounds by the pair; are good layers, but like the Toulouse are rather clumsy sitters, owing to their great weight. The goslings are strong and hardy, but must be kept in an inclosure where there are no weeds and where there is ample shelter from the rain until a few weeks old.

EGYPTIAN GEESE.

Among ornamental water fowl, the Egyptian Geese take a high place. They are a part of the hieroglyphics of the Egyptians, a favorite article of food for the priests,

EGYPTIAN GOOSE.

and their eggs are considered of delicious flavor. As an ornamental water fowl Egyptian Geese are very desirable. They are a rare bird, hard to be obtained, but are easily kept. Their weight is about twelve pounds per pair. They are quite hardy, and by having a suitable pond for them, can be bred as well as any geese. They lay from five to seven eggs at a clutch; and by setting the first laying under a hen, they will lay a second time. They are very pugnacious over their nest and young, and woe to the intruder.

THE CANADA GOOSE.

The plumage of this goose is very handsomely marked; the head a glossy black, as is also the neck, except a band of white across the throat; the upper part of the body is grayish brown; the wing coverts pale gray edged with brown; the lower parts of the body shading into grayish white; the abdomen pure white; and the tail black.

The movements of the Canada Goose upon the land are rather awkward, but upon the water they are extremely graceful, resembling the Gray Swan. The eggs are of a dull greenish tint, eliptical in form, and somewhat larger than the egg of a duck.

CHINESE GEESE.

There are two varieties of the Chinese or Hong Kong Geese, one pale white, the other gray. They are small but excel in egg production. The peculiar knot at the base of the bill makes them at once odd and ornamental. They are more swan-like in shape and carriage than any other known variety.

DUCKS.

Ducks are easily hatched, and easily raised—much more so than chickens or turkeys. Probably the worst thing for ducklings is an unlimited range and water to swim in. The little things are, in a measure, nude, and should be kept in pens with dry soil floors or stone pavement that can be washed down daily. No kind of poultry will succeed on bare boards. All the water they need is best furnished by burying an old pot in the ground and laying a round piece of board on top of the water with room for the ducks to stick their heads in and fish out the corn that is put in the water. This amuses them and does no harm, while, if allowed to go off to ponds or streams, they are very liable to fall a prey to vermin in some shape, or to get their bodies wet and chilled, from remaining too long in the water.

After they are grown, however, many of the varieties must have water which they can swim about in, and also have a reasonable amount of liberty. Those who live near running streams, or who have a lake or pond in close proximity to them, have the matter settled favorably; though perhaps a little more may require to be done with a stream, if it be but a shallow one. Even a little stream can be made to do a great deal in this way by damming it up, and thus creating an artificial lake; or the bed of the stream, for a short distance, may be dug out, and extended a little on each side, which will then make a good place for the ducks to disport in.

Good results uniformly attend the hatching of duck

eggs, when they are from healthy ducks, not in-bred, which are given their liberty and have plenty of water to disport themselves in.

Ducks are excellent foragers. They are incessantly busy in any meadow or pond until their crops are filled. It is a beautiful sight to see them deploy in long lines, running their long bills through the grass in search of snails, crickets and other insects. With a good range, and access to tide-water, they will require very little feed to keep them in good condition.

Before concluding our remarks on ducks, we ought, perhaps, to mention that good, and even necessary, as is the custom of hatching some early ducklings under hens, still it is generally allowed to be unwise to keep birds so hatched as stock birds.

It is also advisable to shut up drakes, or most of them, when ducks are sitting or have young, as about this time many of the drakes are very troublesome, both to mothers and young.

The sexes are soon to be distinguished by their cry, that of the duck being a more decided and quickly repeated "quack," whilst there is nothing like hoarseness from the throat of the drake. The curly tail is not an infallible and trustworthy sign, as we have known many old ducks with a most perfect curl.

The easiest and most common way of judging when good ducklings are fit to die, is to observe if they are "getting cross-winged." They will do well then, and not much before then.

PEKIN DUCKS.

Pekin Ducks are probably the most valuable breed of ducks known to-day. They are very large, mature early and have snow-white plumage. The eggs hatch from two to three days sooner than other varieties, and the ducklings seem larger and stronger at birth. They

can be raised in any place where chickens can, and do not need any more water than land fowls until they are two or three months old. They are excellent foragers and excellent layers. With a good range they require very little feeding. The ducklings can be marketed in July and August, and at this season command high prices. Fourteen or sixteen pounds per pair is not an uncommon weight. As egg producers they are wonderful, being as near perpetual layers as any of the gallinaceous breeds of fowls that can be named.

AYLESBURY DUCKS.

The Aylesbury Ducks are of English origin, and are among native breeds, what the Dorkings are among the common fowls. They are hardy, good layers and good for the table. It is more closely feathered than the Pekin and therefore looks smaller but are, in reality, quite as heavy, if not heavier; eighteen pounds per pair being no uncommon weight, sometimes even weighing fourteen pounds per pair, at three months old. No duck is heavier at that age. For foraging and ranging fields and streams for insect food, the Aylesbury is "ahead." The young will turn out at three o'clock in the morning, in midsummer, and start for the fields to gorge themselves with grass-hoppers that are not as active thus early; for the chill of the night and the dews, making them an easy prey to the ducklings. The leading market duck in England is the Aylesbury. Like other kinds of large ducks it only needs water for drink, and a puddle or place to swim in is not necessary to its highest thrift and development.

ROUEN DUCKS.

The history of the origin, not only of nearly all the various species of our domestic animals, but also of the varieties into which they are divided, is extremely obscure or wanting altogether. The origin of the Rouen Duck is, however, quite certain. The French city, whose name the variety bears, and the district adjoining had but little, comparatively, to do with its "make-up"; but the combined labors of breeders in France and in England evolved, in process of time, from the common ducks, by selection on the basis of size.

Rouen Ducks are very heavy and handsome. They have broad breasts, long, slender necks and long bills.

The Rouen does better than any other except the **Pekin** Duck, where water is scarce.

CAYUGA DUCKS.

This variety is of solid metalic black plumage; is among the most prolific layers. They are said to have originated in the lake whose name they bear; are very hardy and easily reared. Their flesh has a slightly gamey flavor, and by most people it is highly esteemed on this account. They fatten easily and quickly. A pond of water or a running stream is necessary to secure the best results with this variety of duck; as, unlike the Pekin, they will not do well unless they have access to water.

MUSCOVY DUCKS.

Muscovy Ducks are rather odd-looking to those **who**

have never seen them before. Their long tails, their fierce-looking red heads and faces, their oddly blotched and patched bodies (in the colored variety), the oddly ruffled feathers on the back of the drake's neck, and their goose-like hissing, all go to make them objects of curiosity. They have a sort of musky scent, hence the name Musk duck, or Muscovy. They are all good eating at almost any age, if properly cooked, and the young are always a palatable dish. Their management does not differ materially from that of other ducks, save that they require a great quantity of fresh grass, which they should be allowed to pick for themselves, the same as geese; and that they breed only in pairs. An odd, or unmated duck, is useless as far as breeding is concerned, The eggs of the Musk Duck are five weeks in hatching. instead of four, as is the case with the various other breeds common to poultry fanciers. They are useful as curiosities, and also as a practical fowl, to a limited extent.

CALL DUCKS.

There are two kinds of this variety, the gray and the white. Both are quite small and are bred usually for ornamental purposes. The Call Ducks are very pretty, and first-class eating, too. A slice of one of the little fellows goes down very kindly, as any one trying it will admit.

These ducks are sometimes used by sportsmen to decoy wild ducks.

BLACK EAST INDIAN DUCKS.

This is a great favorite as an ornamental water fowl.

The plumage is black, tinted with green. Its flesh is highly esteemed by epicures.

THE WOOD DUCK.

It is quite possible to domesticate the Wood Duck by setting the eggs under a hen, and using great care to tame the ducklings, but for all practical purposes it is not worth while.

CHILIAN DUCKS.

These ducks are very beautiful in appearance. They belong, as a breed, to the aristocracy of ducks; their forms being elegant and their long, slender bodies presenting great elegance in motion. This duck may also be domesticated and is a decided addition to our yards.

Ducks should have a house by themselves, well bedded down with straw or ferns every night. More than four or five females should not be allowed to a single drake, and I prefer to keep them in pairs until sitting time comes, when I kill and use the superfluous drakes.

MANDARIN DUCK.

The Mandarin Duck, sometimes called the Chinese Teal, is the most gorgeous in appearance of all the ornamental duck tribes, and deserves greater attention from breeders and exhibitors than it has heretofore received. It nearly resembles the Carolina or Wood Duck in many respects. It is comparatively small in size. It is closely and neatly built, and the plumage of the

MANDARIN DUCK.

drake in grandeur is almost beyond description. The long crest on the head points backward, and can be

raised or lowered at will. The top of the head is black, which color extends down to the nape of the neck, below which is a clearly-defined white line passing over the eye down to the base of the bill. The cheeks and the long pointed feathers of the neck, and two raised wing feathers are of a bright orange-brown. The upper parts of the breast and back are of a glossy black; and the lower, white. The flight feathers are black and white. The tail black but white underneath. The sides of the breast are greenish-orange, with a clear white marginal line. The legs are a deep pink. From the middle of June to the middle of September the drake assumes the color of a duck; which is a dull, olive brown, mottled, and having grayish under parts. In China domestic specimens are highly prized, being considered as striking examples of conjugal fidelity when paired or mated; hence, in that country, it is customary to carry a pair in the wedding procession, which are afterwards presented to the married couple as objects worthy of imitation. Very high prices have been paid for importations of these birds from China.

CRESTED WHITE DUCKS.

These are quite large and are very beautiful, having large topknots. They are pure white in plumage, and lay well. They mature early and are most excellent ducks for table use. They are prolific layers.

CRESTED WHITE DUCKS.

The crest is large and well-balanced on the top of the head; the eyes are large and bright; and the bill and feet are yellow in color. Some of their offspring are very likely to come plain-headed, as the breed has not been bred sufficiently long to reproduce its like to a certainty. These ducks have many qualities that recommend them to the fancier; but for the farmer, the other white breeds (the Pekin and Aylesbury) have more real economic qualities.

TURKEYS.

Turkeys are pre-eminently a wild fowl and require plenty of liberty to become profitable. Where the range is limited, the breeding of Turkeys should not be attempted. It is on a grain or grass farm where they can roam around at will and gather in the scattered grains and grass-hoppers, that they prove so profitable when well cared for and regularly fed.

It is a great mistake to attempt to hatch too early. May is a good time to set the turkey eggs. The first clutch of eggs that the turkey hen lays should be taken away and set under a hen. The hen turkey soon goes to laying again, and she can be left to hatch these out by herself; by this means you get two broods, while, if left to herself, she will rear but one, and possibly few of these as she trails them through too wet grass. Until two months old, the little turkeys must be well protected from dampness and dews which are always fatal to them.

Turkeys will not bear close breeding like hens for they carry a great deal of their wild nature in their blood, and even under domestication they are allowed to indulge in half-wild habits.

There is one thing which should be remembered by every turkey breeder when he sees the young poults clear from the shell, it is to make sure they are free from lice. And, if they are, let him fix a place for them on a sunny knoll or a spot free from dampness. They can easily be confined by placing long boards edgewise and driving a few stakes on each side to support them firmly. Then make a large coop with a tight roof and put in three or four inches of chaff, cut straw, leaves, or dry sawdust on the bottom, to protect them from the cold, damp ground, for a month or more.

There is nothing so fatal to young poults as cold rains, heavy dews, or sleeping in damp places. They require more care at first than chickens, but make rapid growth when young; and should, by care and food, be forced along as quickly as possible that their bodies may keep pace with the quick growing wing and tail feathers.

Those who grow tobacco should always keep a flock of turkeys, as they are very destructive to the large, green worms that do so much damage to that crop. If allowed a range, and fed on grain at night, they can easily be taught to come up at regular hours.

Turkeys, when full grown, are, perhaps, the hardiest variety of poultry we have; and it is a rather strange fact that this hardy variety is among the most tender when young. During the cold winter months the turkeys will generally be found perched up in some tree or on the ridge of the barn and seem to be contented and take it in preference to a lower and warmer perch. However, it would, no doubt, be better if they could be taught to roost in a certain fixed place, sheltered by some building, or that has at least a partial protection from the cold north winds; and thus, during some of the zero weather of winter, the turkeys will be saved from having frost-bitten feet. Yet, these same turkeys that roost on the edge of a barn in the winter time, can not endure even the least careless exposure when young. They are of American origin, and, although domesticated, their natural habits are but little changed, and it is yet their delight to wander off through the woods and be, in a measure, free from all signs of civilization.

BRONZE TURKEYS.

These are the largest breed of turkeys, and because of their size, hardiness, and the richness of their plumage,

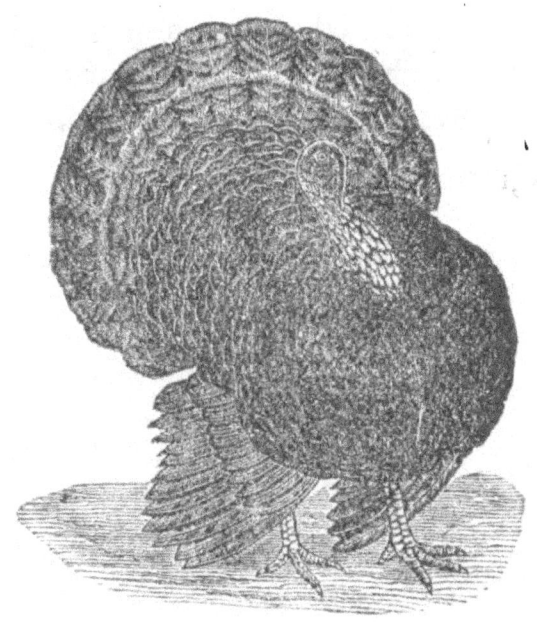

are the most profitable to breed. Their plumage is fairly dazzling when in the sunlight. Gobblers of this breed sometimes reach the immense weight of 40 lbs. The young ones frequently weighing from 15 to 20 pounds. The bronze Turkey is an American bird, of which American breeders may justly be proud. The weight and brilliancy of color can only be kept up by frequent crosses with the genuine wild stock, to which breeders have resource for procuring breeding males every two or three years.

WHITE HOLLAND TURKEY.

The White Holland Turkeys are one of the most beautiful of the turkey family. They are not as large as the Bronze, but are fine sized when well fed. All turkeys are a trifle shy and need to be petted a little in order to have them tame enough to do well, and it is an unexplainable fact that all domesticated fowls and animals when domesticated are more profitable if petted. It is possible that the feeling of security thus given goes a long way towards securing the result, possibly the human magnetism is helpful or healthful in more ways than we know.

The Narraganset Turkey is a bird of black plumage, hardy and of good size.

GUINEA-FOWLS.

The rearing of Guinea-fowls does not offer more difficulty than that of the most hardy chickens. They do not require that particular care so necessary for the successful rearing of young Turkeys. In a state of liberty, free to roam about a large farm, from which they will never attempt to escape, they hatch out their young and provide for them without the help of man. If confined in a run or poultry-yard with other fowls, they never sit, but, on the other hand, they lay an abundance of eggs, which can be given to a broody hen, or put into an incubator. For the first days the young are fed exactly like young chickens. As with the rearing of all the inhabitants of the poultry-yard, they require from time to time a small quantity of finely-minced meat, which gives vigor to the young, and a most robust constitution.

The Guinea-hen is a continuous layer in her season, but does not commence until the grass and weeds get up a little in the spring, as she has a disposition to hide her nest. A little care in observing her habits will soon overcome this inconvenience, as the male bird is always found near the nest, and by his noisy demonstration calls the attention. The eggs of the Guinea fowl are preferred to any other, and the flesh of the bird is considered by many persons superior in flavor to any other domestic fowls, though a little dark-colored. A few hours in salt water will render the flesh quite fair, and as good as pheasant, which it resembles in taste. We consider the Guinea-fowl as a bird that might be raised with profit, besides being ornamental.

WHITE GUINEA-FOWLS.

Young Guinea-fowls reach the adult period earlier than chickens and consequently give less trouble in rearing. At a month old they will care for themselves. A few Guinea-fowls should be kept on every farm, for they never scratch, are a great benefit in the garden and are genuine watchmen, always on the lookout for hawks or other marauders, and will give notice by their alarm cry which can be heard for a mile. The Pearl Guinea is the handsomest of its kind and only for the terrible racket made when they set up a cry would be just the fowl for all village yards.

PHEASANTS.

These beautiful and very ornamental birds are successfully reared in a wire enclosure, The hen lays from six to ten eggs. The old birds will thrive on grain, but the young require an addition of insect food.

SILVER PHEASANTS.

The male Silver Pheasant, without possessing the gorgeous colorations of many species of the group, is a very beautiful bird. The face is entirely covered with a bright vermilion skin, which during the spring becomes excessively brilliant, and is greatly increased in size, so as to almost resemble the comb and wattles of a cock; the flowing crest is blue-black, the bill light green. The upper part of the body is white, penciled with the most delicate tracery of black. The whole under parts are bluish-black; the legs and feet, red; the spurs well developed and usually very sharp. The female is smaller than the male; her general color is brown, mottled with a darker tint; the crest and tail are much less ample than those of the cock; the outer tail feathers are light; marked with black on the outer webs. The female in confinement usually lays from eight to fourteen eggs, and the young are easily reared under a common fowl.

THE GOLDEN PHEASANT.

The male of the Golden Pheasant is in its plumage the most gorgeous of the whole pheasant family, but it is useless here to enter into a minute description of the points of difference.

PEA-FOWLS.

Pea-fowls are the most beautiful of all domesticated fowls, and on large lawns are very ornamental. They will not bear confinement, and thrive only when allowed full liberty. The hens lay two litters a season, of ten or twelve eggs each litter. The eggs should be set under a large motherly hen or turkey. The chicks are very pretty, and quite hardy, much more so than young Turkeys. They will live to a great age, and are very free from the diseases that all other kind of fowls are subject to. As an edible fowl, the Pea-fowl is considered quite equal to the Turkey, the meat having a very gamey flavor.

SWANS.

Besides being ornamental, Swans are often of considerable use in clearing ponds and lakes of all kinds of weeds. The large White Swan is perhaps the most beautiful of all the family. The Black Swan is apt to be bad-tempered. The female Swan lays in February, every other day until seven to nine eggs are laid, and then sits for thirty-five days. More than five cygnets (young swans) are seldom hatched. During incubation it is dangerous to approach the nest which is usually made amongst the weeds at the water's edge, as a blow from the pinions of either male or female may break an arm or leg.

The cygnets are best fed by throwing meal upon the water. The old birds if they are allowed liberty on a large sheet of water, will only need feeding in cold weather; then feed grain. Full-grown males are very quarrelsome and should be kept apart.

PIGEONS.

Pigeons have been considered—a nuisance, I had almost said, but I intended to say, an ornament, and have not entered into fowl literature to any great extent. But a correspondent of the Poultry Keeper gives his experience in

RAISING PIGEONS FOR PROFIT.

William C. Pasco, of South Vineland, Cumberland county, N. J., has for two years been engaged in a novel industry. In the belief that ordinary tame pigeons or doves could be profitably raised in confinement, he built in his back yard a frame sixteen feet square and ten feet high. This he covered on the top and on three sides with wire netting, the meshes of which were about an inch and a half in diameter. On the uncovered side of this cage he fixed a clap-boarded extension eight feet long, with a shed roof. The back of the extension was eight feet high and upon it were fixed about 100 nesting boxes nine inches square and entirely enclosed with the exception of a small doorway in each. A few perches near the upper part of the cage completed the outfit. Mr. Pasco then bought pigeons wherever he could find them and turned them loose in the cage. Some of them pined in confinement and died, others had been injured and did not thrive, but most of the occupants of the cage promptly mated and soon began to multiply. No healthy pigeons were taken from the cage until something over 100 pairs were there in good breeding order. This point was reached eighteen months ago, and since that time the colony has been in prime working order. Mr. Pasco's experience is that each pair of pigeons will raise eight pairs of squabs in a year. The eggs hatch in seventeen days, and when three weeks old the squabs are ready for market, where they have this year sold at something over 50 cents a pair on an average. A careful record of the expense of keeping the birds shows that the feed of a pair of breeding pigeons costs $1.15 a year, and that they will consume food worth 50 cents more in raising eight pairs of squabs to a marketable size. A fair average return per year from a pair of pigeons is put at $2.65, and the net profit at $1.00. Wheat and cracked corn is the regular food, but crushed oyster shells and pounded mortar, with plenty of pure water, are also always kept before the birds.

VARIETIES.

Among the numerous varieties are the Jacobine, Pouter, Tumbler and Carrier Pigeons.

TUMBLER PIGEONS.

The tumbling of the pigeon is a habit which, if seen in a wild bird, would certainly have been called instinctive; more especially if, as has been asserted, it aids these birds in escaping from hawks. There must have been some physical cause which induced the first tumbler to spend its activity in a manner unlike that of any other bird in the world. The behavior of the ground tumbler or Lotan of India, renders it highly probable that in this sub-breed the tumbling is due to some affection of the brain, which has been transmitted from before the year 1600 to the present day. It is only necessary gently to shake these birds, or, in the case of the Kalmi Lotan, to touch them on the neck with a wand, in order to make them begin rolling over backward on the ground. This they continue to do with extraordinary rapidity until they are utterly exhausted, or even, as some say, until they die, unless they are taken up, held in the hands and soothed; they then recover.

HOMING ANTWERP PIGEONS.

The Homing Antwerp is the true "Messenger" or "Carrier" Pigeon, and will return long distances. They are becoming very popular in this country and are in great demand.

DISEASES OF FOWLS.

The three diseases, Parasites, Roup and Cholera, cover most of the ailments of the fowl. Parasites we class among the diseases. Under this head comes Gapes, Body Lice and Mites.

Most of the diseases of fowls are orought about by neglect in the form of exposure to draught of cold air, too intense sun, sudden change of diet, filthy water, food, apartments or close bad air.

If it is decided to doctor the fowl taken with any disease, let it be at once removed to quarters by itself. Before dosing, be fully settled in your own mind what is the trouble, and let it be attended to strictly. It will not do to risk the rest of the flock. If you do not find out what is the matter, and the disease does not seem far advanced, give the fowl comfortable quarters, alone, and with sunlight fresh earth, but nothing to eat or drink, until the question as to what is the matter is settled. Many times this alone will set the fowl right after a few days. Perhaps all she needed was rest, and to give her bowels a rest. Should the disease prove to be cholera or roup, the hand-axe is the only thing that should be called for. I believe the majority of poultry raisers will agree with me when I say, there are more fowls lost by a delay of using the axe than were ever saved by all the medicines known. Never leave the body of dead fowls above ground. This is a sanitary measure that it would seem unnecessary to add, yet what is more common than to see dead hens lying about the alleys in the latter part of the winter.

Right here the old saying "an ounce of preventive is worth a pound of cure" finds its closest application. Keep your fowls well, if possible.

In treating of diseases I shall give as concise and condensed description as possible, together with receipts for all the remedies that I have been able to find.

We will begin with the disease that earliest attacks the chick, namely:

GAPES.

Gapes is a disease caused by the lodgment of small pink worms in the wind-pipe, generally from one to a dozen, and which naturally obstruct the air to and from the lungs, causing suffocation and often death, unless removed. Each worm is double, the male and female together, somewhat in the form of a letter Y, about seven-eighths of an inch long, the male strongly attached to the female by means of a strong membraneus sucker. The heads, when magnified, appear to be all mouth, by which they attach themselves to the lining of the windpipe, with a sucker-like grip. These worms lay eggs in great numbers, which are found attached to their bodies. Just where the gape worm came from, or how or where these eggs are hatched, no one knows.

CAUSE.

Some say the worm is produced from the parasite of the body, which finds its way to the wind-pipe, and takes the form of the worm. One thing writers seem to be agreed on: That chickens raised on old grounds, where chickens years before had the gapes, are subject to them, when on new grounds, or grounds not previously infected with them, were free from them. Foul water, exposure to wet, damp places to brood, and sloppy food lacking nourishment, all tend to produce or aggravate the disease.

The eggs of the gape worm do not seem to lose vitality by freezing or by drying in the fiercest sun, therefore all soil where chickens afflicted with the disease have had their runs, must be *thoroughly* ploughed or spaded under, to prevent future contagion. Just as soon as the chicks are sufficiently old, the muscular tissues of the windpipe become hard and tough, and then there is no more danger of the disease.

SYMPTOMS.

Constant gaping, coughing and sneezing, together with inactivity and loss of appetite.

REMEDIES.

When a chick is suffering with the disease, place it in a close box and burn carbolic acid dry in the box, but be careful that the chick does not become suffocated; still, as he is near death's portals by the disease, it is safe to take a big risk. The fumes from the acid cause the worms to loosen their hold, and become powerless, and the patient will cough them out.

Another. Strip a feather to within half an inch of the tip, dip in coal oil, turpentine or carbolic acid water, push down the windpipe and draw out the worms. Repeat three times.

Burned sulphur fumes are also effective.

A vapor of spirits of turpentine or creosote are recommended.

Sometimes the windpipe is opened from the outside with a knife, the worms removed, and the outer skin sewed up.

Another. Confine the chickens in a small box with a piece of coarse cotton stretched over the top. Upon this place a quantity of finely pulverized lime, and with a stick gently tap the cloth, so that the lime dust will sift through. This will cause sneezing, and the worms will be thrown out in a slimy mass, without danger of being driven further down.

PREVENTATIVE.

Use the Douglass mixture in the drinking water.

Put a bit of the mixture of lard, sulphur and carbolic acid, before described, on the head of each chick as soon as hatched, using care to keep it out of the eyes.

SCALY LEGS.

This disease can be easily cured. The scales are occasioned by myriads of small insects, invisible to the naked eye, but clearly made out by the microscope. They huddle in whitish-gray blotches, at first upon the shanks of the fowls, and if not removed or destroyed early will increase very rapidly. To cure scaly legs, saturate a cloth with coal oil and rub the legs to the shank.

Another: Wash well with strong soap suds, and when dry rub well with a mixture of lard, coal oil and sulphur. Repeat if necessary.

VERMIN.

There are two kinds which feed on the body of the hen: The ordinary hen louse, which sticks close to the body and the hen mite, sometimes called a spider. It is very small, and hardly noticeable, unless looked for sharply. It is of a grey color, excepting when full of blood, when it is red. It is properly the offspring of neglect and filth, and seldom seen in a flock of fowls when at large and not crowded in their houses. They are a hot weather insect or bug, seldom troublesome at any other time. They do not live on the body of the fowl, but go on them, commonly at night while on the roost, to get their food. They are properly the hen bed-bug.

They live in the cracks of the building, on the roost, floors and sides of the building, and in nests. They will gather in large blotches; when full these blotches look red. They will often take possession of adjoining stables, and get on horses and cows, which will rub themselves till they bleed, under the torture of the little pests. It is a very difficult task to get them off a horse or cow.

If fowls have full range they will usually keep themselves pretty free from vermin, but when confined, the keeper must wage a perpetual warfare against the pests.

Whitewashing, with a good addition of sulphur and carbolic acid is perhaps as much of a help as anything. Fumigating with sulphur is also recommended, and tobacco leaves scattered in the nests and about the walls proves helpful. But perhaps the only sure remedy is the Persian Insect Powder, (sold by most druggists). Use with the bellows that accompany the package, mixing four parts of flour to one of powder. Use also on the fowls, sparingly.

LICE.

To be rid of them, provide a dust bath, rub the fowls with Persian insect powder, or what is better blow it in among the feathers with a powder bellows.

ANOTHER REMEDY.

Flour of sulphur 5 lbs: fluid commercial carbolic acid, 1 drachm. Mix thoroughly in a pan with a stick, then hold the hen, put a handful of carbonized sulphur on the fluff and gently work it through the feathers. In five days you can not find an insect with a microscope.

ROUP.

This word covers a multitude of ills. It is often called sore head, sore throat, inflamed eyes, swelled head, canker, catarrh, pustulated nostrils. But it is all roup first and last, in one phase or another. In early stages the hen appears drowsy, goes off in corners of the yard by her self, mopish. She sneezes and gasps for breath; appears choked, and tries to throw something from her throat. After a few days the head swells; offensive watery substance runs from its eyes, nose and mouth; and the comb turns pale or dark colored. The fowl's throat is often so swollen that it can swallow but a little, and in its endeavors it will stand at the drinking vessel, going through the motions for an hour at a time, —the result of which is simply rinsing out its poisonous mouth in the water for other fowls to drink. Hence we see how important it is to remove it at once. Its very breath is poisonous. The disease is contagious and will soon spread through the flock. If the fowl is not a very valuable one, it is best to kill it at once.

I notice that those who advocate keeping a fowl afflicted with roup or cholera, generally have a specific for that disease for sale. I have no medicines to sell. I give the readers of this book the best prescriptions I have been able to collect except such as were copyrighted; those I have not infringed on, but will say that I have secured recipes enough to cure all cases of roup if the rules that I give are followed.

CAUSES.

Exposure to chilling winds, damp floors of houses and damp runs; foul smell of decaying filth: in short, cold and lack of cleanliness.

REMEDIES.

In all cases the fowl should have dry comfortable quarters by itself, and clean water and food.

James L. Bowen, a breeder, gives this as his remedy: "Dissolve a teaspoonful of chlorate of potash in a cup of water, and give a teaspoonful of the mixture once or twice a day. If the nostrils are bad, rinse them with a solution of sulphate of zinc, about the same strength as above, using a small rubber-headed, glass syringe." This he recommends, if the case is taken in time.

Use disinfectants; carbolic acid or copperas in water, sprinkled thoroughly over their perches and runs, and well dried before roosting time.

T. B. Dorsey, another breeder, gives the following as his receipt: "Give one-half teaspoonful to a full one of castor oil, and put in a dry, warm place. Feed with soft food only, mixed with hot ale and red pepper; examine carefully for vermin, and clean. Wash under wings with whiskey. Give little water, and that with tincture of iron or cayenne pepper in it once or twice daily. If no better in two days, give a drop or two of turpentine in morning feed, or one teaspoonful of balsam capaiba. Syringe nostrils, if foul, with weak solution of carbolic-acid water or chlorinated soda. Do not dose too much. Trust the hot ale."

To T. B. Dorsey's rule I will add Kit Clover's opinion which is simply that ale never did any good to man or fowl. I never use it anywhere, and never yet lost a chick with roup. I use an "ounce of preventative" freely.

Another writer says: "Keep them warm, and feed with meal only, mixed with hot ale instead of water. Add Douglas' mixture to the water, and give daily in a pill of meal, half a grain of cayenne pepper, with half a grain of allspice. Give also half a cabbage leaf each day, and wash the head and eyes morning and night, with weak vinegar, or a five grain solution of sulphate of zinc."

Another: Wash the head and nostrils with castile soap suds, weak alum water, or a solution of chlorate of potash, and give cayenne pepper in warm cooked food.

Another: Mix lard stiff with cayenne pepper, and give large doses three times a day. Keep in warm room, and give only warm food.

These all summed up mean: Keep your fowls dry and clean, and housed in bad weather; if then attacked stimulate and clean out the poisonous substance from the head. This is about all that can be done.

CHOLERA.

This is a disease much dreaded in the south and west regarding which there are various opinions. G. M. T. Johnson, a poultry raiser of N. Y. talks learnedly on the subject, and we will quote from him. He says: Until within fifteen years this disease was hardly known in this country. It is now much more prevalent in the South and West than in the North and East. It has been considered by many that it was a disease of filth, spontaneously produced; by others, that certain localities, especially low and damp, were more subject to its ravages, and that cleanliness and dry grounds were the only preventives. I am persuaded that these conditions act very strongly in the operation of this disease, for and against, but from the investigations and discoveries which have been made within the past two years, I am more persuaded that chicken cholera is a contagious disease; that under certain conditions—which are simply taking in the system by mouth or inoculation, one of the poisonous germs—the fowl will be taken with the disease, however well it may be cared for. It has been discovered that the cause of cholera is a microscopic organic germ; that this germ will multiply itself till all of the blood and flesh and excrements are filled with them. It spreads mainly from the excrements. The least particle taken into the system will produce disease. It is also given by inoculation. If with a sharp-pointed knife we scrape off a little skin of a fowl, and touch it with the minutest particle of blood or excrements of a diseased fowl, that fowl will have the disease in regular order, whether the fowls be cleanly kept or not.

I will give some abridged extracts from Professor Pasteur's address before the International Medical College, in London, August 8th, 1881, as to the virulence of this organic germ, producing cholera. He says: Let us take a fowl which is about to die of chicken cholera, and let us dip the end of a glass rod in the blood of the fowl, with the usual precautions. Let us touch, with this charged point, some chicken broth. In a short time the liquid will become turbid, and full of tiny microbes. Take from this vessel as much as can be taken on the point of glass the size of a needle, and touch a fresh quantity of chicken broth, and the same phenomena will be produced. in the same way, the third, fourth to the hundredth, and even the thousandth, —and in a few hours the liquid becomes filled with the same minute organisms. Let us take one of our series —say the hundredth or thousandth—and compare it, with respect to virulence, with the blood of a fowl that has died of chicken cholera. Inoculate under the skin ten fowls, each separately, with a drop of infectious blood, and ten others with a drop of the liquid. Strange to say, the latter will die as quickly, and with the same symptoms as the former ten. The blood of all will be found to contain the same infectious organism. From this the reader will get a good idea of what chicken cholera is and how communicated. The dead fowl should not be buried, but burned, and its house, roosting place and grounds thoroughly disinfected. The best disinfectant is fire, but as that is not practible, the next best is a solution of sulphuric acid. It is discovered that this kills the germ effectually. Sulphuric acid is cheap, but it must be handled carefully until diluted. Make a solution of one pound of sulphuric acid to twelve gallons of water; mix well, and wet the grounds, roost, runs, and everything connected with the disease. Apply with the sprinkling pot thoroughly.

Another writer says of cholera: This disease, which is the most terrible that chicken flesh is heir to, may be brought into the yards, or may arise from filth. The first thing to do when the disease makes its appearance is to disinfect the premises. This may be done by adding an ounce of sulphuric acid (oil of vitriol) to a gallon of water, and sprinkle it freely over the yards, roosts, nests, and the floors. Chloride of lime is also excellent and may be used freely. It will affect the fowls slightly, but will be more beneficial than otherwise. A solution of chloride of lime in water may be used in place of the acid if preferred. The sick fowls should be at once removed from the others. They may be known by evincing a nervous, anxious look, with drooping spirits, great thirst and greenish droppings.

We give also a batch of remedies from contributors to the Poultry Keeper, together with the editor's description of the disease. He says of Cholera:

Cholera may be known by the combs of the fowls changing color from a red to black, and sometimes to a pale tinge; greenish watery discharge from the bowels occur; the bird becomes weak and has great thirst, a nervous, anxious appearance is manifested, and it refuses food. The first thing to do is to separate the sick fowls from the others and give the premises a thorough disinfection, by sprinkling thoroughly with an ounce of sulphuric acid in a gallon of water. The sick bird should be given a heaping teaspoonful of hyposulphate of soda in enough water to dissolve. The medicine will purge the bird violently, and also weaken it. It should then be given five drops tincture of iron, five of paregoric, and one of tincture cayenne pepper twice a day. The feed, should the bird be willing to eat, may consist of boiled rice with heated milk. Boil the rice in water, and heat milk to the boiling point, adding the rice during the boiling of the milk. If the bird continues to be weak and have no appetite, give a powder twice a day composed of one grain of calomel, one grain rhubarb and one grain bread soda.

While we suggest the above we append below the letters from those who give their experience to us, and although we cannot decide which remedy is the best, the reader has an opportunity of comparing them and gaining knowledge thereby.

REMEDIES.

No. 1. Take one gallon of sour milk if you have any if not, something, else that they will eat, and put one handful of salt in it and give it to your chickens. If your chickens are well, take a handful of salt and put it in a gallon of water and set it down where your chickens can drink whenever they feel like it. If you follow these directions I don't think you need be afraid of chicken cholera any more.

No. 2. For any number of fowls one may have, take corn meal one quart, one tablespoonful of saleratus, mix together and wet up with new milk. This I know to be an effectual cure from experience, for where from one to three would be found dead under the roost every morning, one feed of it being fed them, not one died after it.

No. 3. I will give my remedy for cholera in advanced stages with which I saved several turkeys last year. One teaspoonful of laudanum, the same of ground pepper mixed with a little soft food, and given three times a day. The above amount is sufficient for three doses. I just gave it one day and the next day, still poking down their throats, three or four times a day, a little nourishing soft food, after which set free and let them shift for themselves. If they are not too far along with it, this remedy will cure. I think in the case of cholera an ounce of prevention is worth many pounds of cure, with moderate cleanliness and plenty of pure water with a lump of copperas dissolved in it; occasionally some small grain mixed in with corn for food; and broken dishes and glass for them to get at all the time, one need have no fear of the plague whatever.

No. 4. I will give you a sure cure for cholera and also for gapes in young chickens: Two tablespoonfuls of cayenne pepper to one cupful of corn meal mixed up wet, then take some onions and chop them up fine, so they can eat them without trouble; the onions are for gapes; too much can hardly be said in praise of onions for fowls or chicks, they are good for them, and ought to be fed twice a week. Also try oil cake for fowls and notice the result; feed it twice a week broken fine, and inside of thirty-six hours a great change will take place;

the comb and wattle will become of a beautiful scarlet, the eyes brighten, the motions quicken. It should not be fed to young chicks as its effect will be radical on the weak systems. Epsom salts is the best remedy I know of for cholera; mix it with their food once a day. one-half tea-spoonful to one cup of mashed potatoes or corn meal.

No. 5. Here is the ne plus ultra of cholera cures: Take one ounce meal, one-fourth ounce ext. gentian and twenty-five drops of carbolic acid formed into twenty-five pills, and give one pill three times a week.

No. 6. From Jos. Swonger, Springfield, Ohio: In 1880 I bred, imported and raised 500 chickens, had fine prospects for a good sale and nearly all ordered, when the cholera attacked my yard, and in two weeks I lost my entire stock. I tried everything, all kinds of cures and preventives, and they did no good. I went to work and boiled white oak bark, cherry bark and peach leaves together and made a syrup, and set it away till cold, gave it in their drinking water twice a week, and soon my chickens looked fine and their combs became red, and they have never had the cholera since. It is a fine remedy and does not cost anything. Try it. I am never without it. You will never have the dreaded disease.

No. 7. One lb. of copperas, one lb. of ginger, one lb. madder, one lb. soda, one-half lb. of black antimony, one-half lb. cayenne pepper. Mix in meal and feed twice a day for three days. Stop it for two days, and then put some in the drinking water.

The above remedies are all different and show that no

certain cure is known, though any of the above may answer in some flocks. We believe that climate, season and soil have an influence with cholera. A medicine that cures in one locality may fail in another.

In July we lost a whole hatching of young chicks with the cholera, and though I tried all the remedies I could hear or think of they kept dying till we lost 150. Soon after I saw the following remedy: "Spanish brown 1 lb.; black antimony 2 oz.; cayenne 2 oz.; a teaspoonful to a dozen fowls twice a week as a preventive, once a day if sick." I immediately had it prepared, and though our chicks were kept closely confined I did not have another case of cholera till I got out of the medicine and neglected to get some more, when it soon made its appearance again, but I only lost one chicken, and don't think I should have lost that if I had given it the Spanish brown, instead of alum. A few days since I found a Polish Bantam drooping; its crop was full of sour food; breath bad; droppings greenish and frothy. I gave it a little of the medicine in some meal and the next morning the crop was empty and the chick was as well as ever.

I don't know whether it is good for other diseases or not, but since using it, have not had a case of roup or gapes, and have 300 as nice chicks as anyone would care to see.

Poultry books say vaccinate, but I have never seen it done, and therefore am not authority but here is the rule: vaccinate a hen and in eight days her system will be thoroughly inoculated; then cut off her head and catch her blood in some vessel; then pour the blood out on paper to dry. A half drop of this blood is sufficient to vaccinate a hen, and the blood of one hen will vaccinate your whole flock.

Catch the fowl you wish to vaccinate, and with a pin or knife make a little scratch on the thigh (just enough to draw blood); then moisten a little piece of the paper with the blood dried on, and stick it on the chicken's leg where you scratched it; then let the fowl run, and you need have no fear of chicken cholera.

TOBACCO REMEDY.

"Take a piece of plug tobacco and pour hot water over it, making a strong tea; then make your dough up with it, and feed to your chickens three times a day. It has never failed to cure for me, unless my chickens were too sick to eat. When I find they are sick with the disease, I keep the tobacco in the chicken trough. If it is too strong, the chickens won't drink it. The trough should be kept very clean."

Another remedy: Alum two ounces, resin two ounces, copperas two ounces, sac sulphur two ounces, cayenne pepper two ounces: pulverize, and then mix three tablespoonfuls of the powder with one quart of corn meal and dampen for use. This quantity is sufficient for twelve chickens, and may be used either as a preventive or as a cure; for the first it should be given once or twice a week.

Another: Try pulverized nux vomica for chicken cholera; one teaspoonful to twenty hens, twice a week given in food, is a never failing preventive. To all that are visibly affected, give a drench made of the same.

This is a sure cure for chicken cholera, and, as far as I have tried, is equally as good for hog cholera.

Two tablespoons of copperas, and pour one pint of boiling water over it; mix with corn meal and feed it to the fowls.

Red Venetia is said to be a sure cure, and also a good preventive. Mix one tablespoonful with a quart of meal and feed two or three times as a preventive.

Dissolve chlorate of potash and indigo in water or milk. This is a good preventive.

Shower the houses and grounds with one-half per cent. sulphur acid. Use Douglas Mixture. Give blue mass and cayenne pepper, each one ounce, one teaspoonful laudanum, one half ounce gum camphor; mix well and make into pills. Give one pill every hour until purging ceases.

FEATHER PULLING.

This comes from a lack of animal food. The small breeds, being active in habit; are subject to it when in confinement. The best preventive is to supply the fowls with a variety of food, and if meat cannot be procured, use less corn and more bran, wheat, buckwheat, and oats. The vice is not inherent, but acquired. Should one hen in the flock become addicted to it, she will teach the others. There is a little device made purposely for stopping this habit. It is called a "Poultry Bit" and will be sent by mail for 15 cents, two for 25 cents, four for 50 cents, or ten for one dollar.

EGG EATING.

Sometimes hens eat their eggs, but this bad habit may be cured by making the nests dark, leaving only a faint light in them. Hens prefer a secluded darkened nest, but they will always return to the light as soon as they can, and they will leave the egg untouched in preference to eating it in the dark. It is this aversion to darkness that prompts them to prefer the open air in winter, rather than remaining inside.

CROP-BOUND.

Should a fowl become crop-bound, work the crop well with the hand, and endeavor to force away the obstruction in the passage-way to the gizzard. Should this fail, draw the skin to one side and cut the crop sufficiently to relieve it of its contents. Sew up the wound with silk, and the fowl will not be seriously damaged. After cutting, be sure that the obstruction in the passage is removed as well as the contents.

DIARRHEA.

Caused by stale food or filthy water, exposure to hot sun, or foul or crowded quarters. Renovate house and yards and give Douglas Mixture freely. Use little food and no corn.

SOFT SHELL.

Give plenty of lime, pounded bone or oyster shell broken fine.

FROSTED COMB.

Mix two parts of glycerine and one part turpentine; rub the affected parts every morning with the mixture.

At noon apply a compound of three parts sweet oil and one part rose water, and at night apply the first mixture again as before. A few days of this treatment will be pretty sure to restore the parts to their normal condition.

BUMBLE FOOT.

This is an unsightly excresence, which is very apt to appear on the Asiatics and other heavy breeds. It seems to act very much as does a stone bruise on a boys foot. An application of lunar caustic or iodine to the surface may sometimes reduce the swelling. Should suppuration set in, open the wound, and if possible keep out the dirt and wet long enough to give it an opportunity to heal.

LEG WEAKNESS.

Leg weakness is nothing but the result of high feeding, rapid growth and forcing. It is not necessarily fatal. Mix a little bone meal with the food, put a teaspoonful of copperas in the drinking water; and with a variety of food the chicks will get over the difficulty without trouble.

APOPLEXY.

Caused by over feeding. Bury the dead fowls and feed less.

WORMS.

A good dose of castor oil followed by a little sulphur mixed with the food.

And now I will give you a few recipes that are sold every day at fifty cents each. First in importance, is the celebrated

DOUGLAS MIXTURE.

Take of sulphate of iron (common copperas) 8 ounces, sulphuric acid $\frac{1}{2}$ fluid ounce, put into a bottle or jug one gallon of water, into this put the sulphate of iron. As soon as the iron is dissolved add the acid, and when it is clear the mixture is ready for use. This mixture is to be given in the drinking water every two or three days. A gill for every 25 head, or one teaspoonful to a pint of water. If there are any symptoms of disease it should be given every day. This is one of the best tonics known to the poultry fraternity.

THE CELEBRATED BUCKEYE EGG FOOD.

Buckwheat, 8 quarts, Indian corn well parched, 8 quarts; oil cake or meal, 8 quarts; oats, well parched, 8 quarts; Egytian rice corn, or wheat, 8 quarts. Grind all together, then mix the oil cake in and add one pint slacked lime, 1 pint of ground bone, 6 tablespoonsful of common salt, 5 tablespoonsful capsicum. Put all the above ingredients together and thoroughly mix. This will make about two bushels of feed after being ground. Cook as much of this feed as your hens will eat at one time. and feed it in the morning, warm. Do not put in too much water that the feed will be sloppy, but have it dry enough that when thrown down it will not break apart. Feed this food two or three times a week, and not oftener, for remember the food is very strong.

ROUP PILLS.

Equal parts of ground saffron, asafœtida and hyposulphite of soda, made into pills the size of a pea; or given in as much powder as will lie on a cent, twice a day to each fowl.

THE HAVANA METHOD

Has already been given. We will give the experience of a large shipper. He says regarding the "Havana Method." "They keep perfectly, but have a dark, soiled appearance, and mostly rough shells, which easily distinguish them from fresh eggs, though I think this is owing to the alum in solution (which I shall leave out next season). My formula, excepting the salt and lime, is: Five ounces baking soda, five ounces cream tartar, five ounces saltpetre, five ounces borax, one ounce alum, to twenty gallons lime water. This is for a whiskey barrel full or about one hundred and fifty dozen eggs."

POULTRY CONDITION POWDERS.

Pulverized Ginger........................1 pound.
    "   Flax Seed......................1   "
    "   Licorice Root..................1   "
    "   Blood Root....................1   "

MISCELLANEOUS HINTS.

TONICS.

Lewis Wright in his Illustrated Book of Poultry, recommends the following tonics:

NUMBER ONE.

Licorice............2 oz
Ginger.............2 oz
Cayenne Pepper....1 oz
Anise Seed.........1½ oz
Pimento............2 oz
Sulphate of Iron....1 oz
Powder and Mix.

NUMBER TWO.

Cassia Bark........1½ oz.
Ginger ....5 oz.
Gentian............½ oz.
Anise Seed.........½ oz.
Carbonate of Iron..2 oz.
Powder and Mix.

NUMBER THREE.

Peruvian Bark.....2 oz.
Citrate of Iron.....1 oz.
Pimento..........2 oz.
Cayenne Pepper....1 oz.
Gentian...........1 oz.
Powder and Mix.

NUMBER FOUR.

Cascarilla Bark....2 oz.
Anise Seed........½ oz.
Pimento...........1 oz.
Malt Dust.........2 oz.
Carbonate of Iron..1 oz.
Powder and Mix.

No. 1—Best for sudden colds. No. 2—For cold and wet weather, and young turkeys. No. 3—A restorative after long journeys, exhibitions, etc. No. 4—Where a continuous use of tonic is required, for general debility and the like. Only enough of either should be used to give the food a slight characteristic taste.

BILL OF FARE FOR LAYING HENS.

The shell of an egg consists chiefly of carbonate of lime, similar to chalk with a very small quantity of phosphate of lime and animal mucus. The white of an egg—albumen—is composed of eight parts of water; fifteen and a half parts of albumen and four and a half parts of mucus, besides giving traces of soda, benzoin acid sulphurated hydrogen gas. The yolk consists of water, oil, albumen and gelatine. Now hens must have something to form shell; oyster shells head the list, bones of any kind are good and by roasting or burning them until they are brown and brittle you have almost the genuine egg-shell; lime with gravel and sand is good. Albumen—the white of the egg—is found almost in its pure state in fresh sweet milk, and in wheat, oats, rye and buckwheat—barley and corn in the order named, corn with other grain furnishes oil and gelatine. While at large and during the summer season hens get plenty of seeds, weeds, etc., that furnish a great portion of the items named. The bones and shells they are not apt to get, and they seldom ever can find albumen enough. Now this makes plain what we are to feed.

FOR BREAKFAST.

Take one part meal or cracked corn, one part shorts and one part bran, or one part meal, one part chopped oats and rye and one part bran, mix with milk, or water if in the winter season—the milk or water should be boiling hot. Do not use enough milk or water to make the mess sloppy. Occasionally season with red pepper. One or two mornings during the week instead of the above, feed "Buckeye Egg Food."

FOR DINNER.

Feed either wheat or wheat screenings, oats, rye, buckwheat or barley, with the preference in favor of buckwheat or wheat.

FOR SUPPER.

Parched corn.

During the winter give occasional feeds of boiled vegetables, and meat scraps, or cracklings. Beef's liver or lights are good, and can be bought cheap.

HOW TO FEED POULTRY.

Attention must be paid to this part of the industry, in order to make it profitable, and if you want to keep the fowls thrifty and in good health.

While it is most essential to be careful not to overfeed, the extreme of underfeeding must be avoided. It isn't the gross amount of feed they get, but the mode in which it is distributed to them that tells to the best advantage.

A variety of diet for fowls or stock is a very important matter to be remembered by poultry-keepers, both on the score of economy and for the best good of the fowls themselves.

Never feed more than the fowls will eat up clean. It is better not to feed enough than to feed too much. Through the summer and fall months Douglass Mixture should be given in their drinking water.

For this reason I deem it important to frequently remind those who are endeavoring to breed poultry to give profit, no matter whether for fancy or market purposes, that a fixed system should be adopted and adhered to continuously, if we would make the most of the food given the fowls and chicks. It costs no more, really, to afford them variety than it does to stint to a single kind. I earnestly advise that the widest range be given to such varied feeding, whenever it can be done conveniently.

HOW MUCH FEED FOR A HEN.

The rapid digestion of food in a fowl's stomach calls for an almost continuous supply of food. In a discussion of this question in the *New York Times*, that paper says: Regularity of supply is of the greatest importance when the consumption is so rapid. If the supply is not regular, there is great loss. There is not only a waste of energy, but a waste of time in restoring this waste of power, and it is on this account that so many fowls merely live along and do not produce eggs as freely as is expected, although the quantity of food is supposed to be quite liberal. The system of sending the fowls to roost without a full belly, and keeping them all hungering and thirsting for food, is to blame for much of the short-comings of the hens. The quantity of food required by an animal is estimated at about three per cent. of the live weight daily. This merely supports life; all increase of weight, or any product whatever, must be supplied by an extra allowance, so that twenty hens, weighing 100 pounds, would need three pints of solid nutritious food daily to live, and do no more. This is equal to three twentieths of a pint for a hen. Two twentieths, or one-tenth of a pint, or about one and one-half ounces of food, is then required every day for the production of eggs, the total daily food requirements being one-fourth of a pint, and this is the established rule, from long experience, among poultry keepers. One quart of corn or other grain for eight hens is the regular daily allowance, given in at least two meals, and it has been found that a flock of hens, when supplied with a constant provision of grain before them, will consume this quantity and no more, in addition to what small things in the shape of flies and other insects, grass, etc., they may pick up.

When fowls are kept too cold it takes all they can digest merely to keep up animal heat. Often the same quantity of food would be sufficient, if they were kept comfortably warm, to induce laying.

Mr. H. Blanchard, California, gives some figures of experience and profit in *The Poultry Keeper*, which we append:

Last year I kept 100 hens. I fed two tons barley, $40; two tons middlings, $40; two tons shorts, besides stuff from the garden, $40; total, $120. Sold 400 dozen eggs at twenty-five cents per dozen, $100; set 4,000 eggs, hatched seventy-five per cent.; lost in rearing, ten per cent.; balance at three months old, 2,600. Sold in San Francisco at thirty cents per bird; commission and freight off five per cent. fifteen cents for raising, net ten cents or $260. You will notice that old hens paid for their food for the year, leaving $260 as a net profit. These fowls were confined. I used no incubator but kept thirty head of turkeys that hatched out three broods, without leaving the nest—nine weeks. I fed a fraction over one and one-fifth pounds per head, a day.

There is nothing better for poultry than milk, one writer says of it:

"A neighbor of ours whose hens, to our exasperation, kept laying on when eggs were forty-five cents a dozen, while ours persistently laid off during the same season, on being questioned revealed the fact that his had a pail of skimmed, perhaps clabbered, milk each day, and no other drink. On comparing notes we each found that our management of our fowls was almost exactly alike with this single difference—a difference that put many a dollar to the credit side of his ledger, while our own was left blank during the same period; and this thing had been going on for years, with the result always in favor of the milk diet."

It is estimated that a ten-weeks old chick will cost ten cents.

Poultry need lime with their food. The common food alone will not furnish lime enough for a full supply of eggs. In a state of nature a hen would lay a single litter of eggs, hatch them, rear the chicks, and then give up business for the season. The ordinary food would supply this small demand. But when a hen lays 120 eggs she will want as much lime in a month as she would naturally get in a year. This excess must be supplied. Crushed bone and oyster shells are the best, and should be kept always within the reach of the hens. It is not advisable to give egg shells unless they are broken up very fine, otherwise the hens may learn to break and eat eggs.

SUNFLOWER SEEDS.

I have found sunflower seed to be one of the best of feeds to stimulate egg production (it should not be used too freely as the seed is very rich in oil; indeed, it is so rich it can be made to take the place of meat or beef scraps, to a great extent), being much superior to many of the so-called "egg foods," that have been so widely advertised all over the country, which if fed to your fowls frequently do more harm than good. It also gives that beautiful rich gloss to the plumage which can be produced in no other way, an item which every amateur should know, who ever expects to exhibit a bird at a poultry show. Sunflowers are also useful in other ways.

When planted in low, damp, or ill-drained localities, they absorb miasma, thus preventing fevers. Honey-bees gather large quantities of pollen from the blossoms also. The common single variety has heretofore been much grown for seed. But the new mammoth Russian variety is so much superior in every respect, being of a stronger, more vigorous growth, having larger heads which on moderately rich ground will average from ten to twelve inches in diameter and frequently larger, the seed being also nearly twice the size of the common sort and in color white with black stripes, that it is now quite generally grown by our more progressive poultrymen. Plant in the back yard or in corners of fences if no other spot can be spared.

It is estimated that one bushel of corn, or its equivalent, will keep a hen a year, but they no doubt require more when in confinement. Much depends upon the breed, climate, health and productiveness of the fowls. We think five pecks a better estimate for a year's feed Cottonseed meal should be fed only twice a week, and only a teaspoonful for each hen at a time. Linseed meal is better.

The *New England Farmer* states that the coops for broody hens at Houghton Farm, are about two feet square, with bottoms of slats so small and far apart that no hen, however broody, will imagine she is sitting while roosting upon them. They are raised about four inches above the ground. It seems to be a first-rate contrivance for breaking up sitters.

If a spoonful of fine bone meal be mixed with the soft food for a dozen chicks three times a week it will be sufficient, but the quantity should be gradually increased as the chicks grow, or the same quantity may be fed oftener.

TO FIT FOWLS FOR EXHIBITION.

When competing for prizes the fowls are put in light and tasty coops, in pairs and trios, as the Society may direct. The coops are generally open at the top and in front, with exception of slats; and a slide-slat or two in front, by which to put fowls in and out. This is to let as much light in the coop as possible, that the fowls may show off well. Of course the first point is, to get superior fowls. The next is, to have them appear well. The cock and hen should match in the pen—that is, be of the same shade of coloring, of proportionate size, and about the same age. They should be clean, and their feathers smooth, none pulled out, especially such prominent ones as tail or wing feathers. Many a first-class fowl has given up his medal to his inferior, for the reason that he did not show up well. They should be somewhat used to the coop, and not wild, that when the judge takes them out and handles them, he can get at their true merits, and that they may act natural.

Fowls designed for exhibition should be put in clean and comfortable apartments by themselves, where they will be quiet for a week or two before exhibition. If necessary, they should be washed, and allowed good time to dry, in clean apartments. For feed, whole grain is the best. Let them be in good order, and not over-fat. Sunflower seeds are good to feed them, with other grain, for two weeks before exhibition, but they have to learn to eat them. If sunflower seeds are not convenient, give the hens oil meal or cotton-seed meal, mixed with Indian meal and wheat middlings, in the proportion of one-fourth oil meal to three-fourths other meal, once a day. This tends to make them vigorous, active, and hold up their heads; feathers shine and lay smooth. The judge will pass upon every part of the fowl—head, comb, legs, wings, tail, condition, symmetry, etc. A perfect fowl is allowed 100 points of excellence. A certain number of points will be allowed to the comb, say five; a certain number, say seven, to wings; and a certain number, say ten, to legs. (With different kinds of fowls these numbers are proportioned differently.) These whole numbers added together make 100. If the comb is defective,—perhaps ill-shaped, twisted or lopped,—one, two or three points will be taken off, according to the seriousness of the defect; so with all parts. The points of excellence left for each fowl are then added up, and the one whose number is the highest ranks best. I have never yet heard of a fowl that possessed 100 points. Many good fowls fall below ninety. It is a superior fowl that will score ninety-five points.

Fowls of some varieties, that will score ninety-eight and ninety-nine points are sold, now-a-days, for $75 and $100 apiece. There is need of great care, that the changes of condition of the fowl to the coop, and at liberty, do not work against his health. Many a good fowl is lost just after exhibition.

WASHING EXHIBITION FOWLS.

The washing of white or light colored fowls that are to be sent to the shows has been practiced for a number of years. It seems, however, not to have been a rule with breeders of other varieties, light or parti-colored, if not white, to wash their birds before shipping them, though such a preparation certainly does no harm. Washing a bird is a very simple thing to do, but it takes a little care and knowledge to do the thing properly. Any one can wash a fowl, but a green hand going simply by rule of thumb is likely to turn out bad work. The birds should be taken into a clean, warm room, and there washed one at a time. The only things needed are a sponge, soap, warm water, and some clean pieces of old flannel or blanket. The bird should be placed in a tub of blood-warm water, and pressed down gently till its back is level with the water. A clear soap suds should then be made and rubbed thoroughly into the skin by running the sponge in the direction of the feathers, working clear down to the feet. Great care must be taken not to break the web of the feathers, and under no circumstances should the sponge be rubbed up the feathers. When this has been done the fowl is taken out and rinsed in cold water to remove the soap, and then carefully dried with the flannel.

Some breeders wrap the fowls in cloths to prevent their moving until their plumage is completely dry, while others give them their liberty in the room as soon as they are rubbed down. Care must be taken that they are not exposed to cold air till completely dried. The shanks, feet, comb, and wattles are washed in rum or alcohol and water to produce a clear, healthy appearance, and the birds will then appear in the show room at their best.

SHIPPING BOXES.

The *Poultry World* says that a box in which a trio

or quartet of full grown fowls are confined in a journey need not be larger than 24x18x18 inches. The material for this box may be half inch stuff. The front and ends can be open lathed, and the back of unbleached, stout cotton. The bottom and top of whole boarding will be strong enough. In cold weather stretch the cloth nearly around the entire front and ends to prevent the freezing of the birds combs. Feed sufficient to afford the birds half a pint each per day of whole corn and wheat for the term they may be en-route, and a common tin pint cup for drink, will be all that is necessary for their convenience. In the bottom of the box strew a layer of hay or short straw, and the whole will weigh but twenty pounds or so, in addition to the contents. If the above plan, in a general way, be adopted by shippers, the cost of transportation to buyers is lessened, over the careless method too often practiced of sending fowls in a heavy, solid inch-board box, that weighs more than do the birds themselves.

The back of an exhibition coop should never be made of slats, for birds placed in such a coop will, at the approach of a visitor, turn their heads toward the open work back and try to escape thereat, thus presenting themselves to view in the most unfavorable position; but if the back of the coop is without openings, the birds see no prospect of escape, and therefore face the front of the coop, in which position they show off to the best advantage.

PACKING EGGS FOR HATCHING.

Take a light flat-bottomed splint basket, large enough to hold the number of eggs you intend shipping, in one layer; lay a folded newspaper in the bottom and let it extend up the sides; then put about two inches of hay chaff on that; then take paper about eight inches square and wrap each egg separately, twisting the paper well at each end; set them, large end down, in the chaff, so they don't quite touch each other, then put well-dried sawdust between them, press it down gently so as not to break the eggs, then put hay chaff to come up about an inch above the eggs; then lay a piece of paper over and cut a piece of muslin a little larger than your basket; lay it over the paper smoothly, thread your darning needle with wrapping twine and sew the muslin smoothly, passing the needle through the basket and drawing it well down so that in jolting the eggs won't move around, then sew your shipping card on top of your muslin, and your basket is ready to start.

NUMBER OF FOWLS TO THE ACRE.

First, do not mass your fowls. If you have but one acre of land to devote to poultry, divide it into eighths, and put not more than twenty-five to a flock, which will make room for 200 fowls. The same acre will not keep fifty if allowed to run together, as well as the larger numbers separated. In fact 200 fowls will hardly flourish on five acres if kept *en masse*. Let your motto be small flocks and complete separation.

It is not natural for fowls to run together in great numbers, promiscuously. When a great number of fowls are purchased from different sources and brought together they are afraid of each other and fight almost constantly for a number of days, or till they become acquainted. When a flock is of the proper size, each individual is acquainted with every other; just as well as a scholar in school knows every one in his class. Also every fowl in a smaller flock knows the relative strength and courage of the rest. There is one that is the "boss," able to beat any one of the others. Then there is number two, number three, and so on. When they become all settled down, and each one knows its place there is order and quiet. Without this order and quiet there can be no thrift. All gallinaceous birds live in this way, in a state of nature, the wild jungle fowls, the parent stock of our domestic fowls, included. Each family group has, by tacit agreement, a certain district for its beat. If a member of any group or family strays over the line it is regarded as an intruder and driven back.

If we substitute for this state of things, a mob, or heterogeneous assemblage, the fowls are kept in a continual state of worry. The hereditary family instincts are violated, and the laying is checked.

On a village street where the houses are ten or twelve rods apart, each resident can keep a flock of thirty fowls or so, and when there are no fresh purchases made, and the birds have all settled down to business, they stay at home and rarely intrude on the range of neighboring flocks.

We receive a great many questions in the course of a year, regarding the proper allowance of ground, for a flock of twenty, or fifty, or 100 fowls. To begin with, there should be no flock of 100 fowls at all. Twenty to thirty is the maximum number; perhaps in some cases fifty. Now how many fowls to the acre? Well what is the object to be attained? Is it simply to afford place to run around in; or is it to give a chance to pick up something to eat? If the latter, does the poultry keeper aim to secure vegetable forage or insect forage, or both?

One thing is certain if it is expected that fowls will pick up insect food, as well as vegetable food, it will take four or five acres, to give good foraging ground, for three or four flocks, of twenty birds each. Grasshoppers and crickets and various other insects need space, a variety of vegetation, in order to get a living, so that they may be in turn consumed by the fowls. If there is a large pasture used for horses or cattle, and but a small number of fowls, it may be profitable to raise the animal food, as well as the vegetable food for your poultry. In this case they can have the range of five or ten acres or more. They will thrive exceedingly well under such circumstances. Leghorns especially may frequently be found a fourth of a mile or more away from the fowl-house. A flock of Leghorns under such circumstances, will make use of a range of over eighty acres.

But the majority of our readers can not allow any such extensive range for their poultry. It is not to be expected, ordinarily, fowls will be able to procure any considerable portion of their subsistence, by feeding upon insects. Now if we give up as impracticable, the foraging for insects, and aim to allow sufficient range for vegetable forage only, we shall still use considerable ground, though much less than would be required if it was expected that fowls would hunt grasshoppers and crickets and other things of this sort for their living. The reason why considerable ground is necessary in order to raise grass is that fowls trample down much more than they eat, and they will wear out their range so that it will become completely bare. Perhaps it is the cheapest way, where land is expensive, to cut the vegetable food and feed to the fowls.

But give all the range possible to the fowls unless obliged to confine closely, in that case allow a yard thirty feet square for a flock of twenty fowls.

Regarding the subject a writer in the *Poultry Keeper* says: The number of fowls that may be kept on an acre of ground depends upon the management given them. An acre is about 209 feet square and can be cut up into twenty yards 20x100 feet each. Estimating a cock and ten hens to each yard, we can consequently accommodate 200 hens. But to say how much each hen should produce is not easily done, for everything relating to their ability to produce a profit is to be considered, such as the breeds to which they belong, their ages, the character of the soil, and climate.

But if one intended to use an acre of ground only for poultry, a large revenue may be derived from chicks. They should be hatched in incubators, and, as the earlier broods will constantly be removed in order to give place to succeeding lots, the actual space required for a large number of chicks is very much lessened. Ezra Packard, of Hammonton, N. J., once kept over 1,000 on a piece of ground not over fifty feet square, and the chicks were not sold until they weighed several pounds each, owing to the fact that he did not hatch them till late in the season. He kept them in brooders dotted here and there over the yard, and had but a small percentage of loss. The proper method, therefore, would be to devote one-half of an acre to hens, in order to procure eggs suitable for hatching, and use the remaining for broods. If they are hatched early and marketed as soon as they arrive at a suitable period, the number that may be kept is almost beyond estimate. Had Mr. Packard's lot of 1,000, which he could not have secured with hens, been hatched early in the season his small piece of ground would have netted him hundreds of dollars, and he was not restricted to space, as he resides on a small farm, but it became apparent to him that he could do much better with them on a small area, than when they are scattered over a large field.

With the use of incubators no great space is required. An acre may produce $2,000. The result depends upon the season, the breeds and the management, and every hour of labor so devoted will be amply rewarded.

But what is the profit from hens alone may be inquired. There is no sum that can be stated. Many records of what the hens pay are published every month for the benefit of those interested, the profits ranging from $1 to $7 per hen, but the profit depends upon the whether the breed used is a large or small one; whether the fowls are non-sitters; whether they are good winter layers, and whether they are well managed. The location of the place, climate, markets, soil, and prices obtained are also very important factors. The fewer the number in each family of fowls the better. Dividing the flock is an advantage always, as no large number have ever been kept together yet without some contagious disease sweeping them away.

POULTRY ON A LARGE SCALE.

I have not in this little book attempted to speak of poultry raising in large numbers, my suggestions applying in the main, to the treatment of small flocks of fowls such as are kept on the ordinary homestead.

If it is desired to rear hundreds, instead of dozens, greater space must be afforded the several flocks; and proportionate runs, houses, and conveniences, must be provided for the needs of these increased and increasing numbers. Any one can advantageously manage forty or fifty adult fowls and chickens, who will follow the directions we here publish. But if hundreds of fowls are to be bred, the work becomes more complicated; and the results are not always so fortunate as is anticipated, when this is undertaken by inexperienced parties.

To attempt the keeping of a thousand or more fowls upon one place for instance, is not the child's play that some persons seem to fancy it to be. In a single body or collection, no such numbers *can* be kept together with profit. Plans have been proposed, and there are writers on poultry, who tell us how this may be accomplished to advantage. But this thing, like many another problem in successful fowl-raising, is not yet solved.

I do not in this assertion assume that one thousand or ten thousand domestic fowls can not be managed upon one estate (provided the farm be large enough), and under one competent general superintendent. But what we intend to convey is this: to keep large numbers of fowls upon one place, the flocks must be colonized, with not over thirty or forty or so together. For each colony separate houses must be provided, and ordinarily these runs must be fenced, and to keep them in good thrift throughout the year, each lot must have ample space for range.

This requires a great deal of land and it also requires so much attendance to feed and look after this excess of numbers, that the cost of their care, feeding, doctoring, housing, etc., will eat up the income that can be realized from them, unless there is first rate skill and ability employed.

Poultry on a large scale is simply a combination of many houses and yards instead of a few, with the exception of larger mills for grinding food, and larger boilers for cooking it. A person who *thoroughly* understands caring for 100 chickens, is competent to care for 1000, provided he has plenty of room for houses and yards.

THE TURKEY AS AN INCUBATOR.

Here is what one writer says on the subject:

I have been experimenting in a small way with the turkey as an incubator, and am very much elated over my success. I find that a turkey placed in a cage slightly darkened and supplied with a nest of porcelain or glass eggs, will in from four to six days become cross and broody. I then remove the porcelain and give twenty-five or thirty hens' eggs. Of these twenty-five or thirty eggs she will hatch nearly all the fertile eggs. The turkey seldom breaks an egg, and if allowed to mother the chicks nothing can equal her as a mother.

A friend has been using this mode of incubation for three years past. He thinks nothing can equal it. He uses from twenty to thirty turkey hens every year. He places them in the cages about January 15. In four to six days they are ready to set. He keeps them at work about nine weeks, or long enough to hatch three broods each, or about seventy-five chicks each. The young chickens are removed when about twenty-four hours old and placed under artificial mothers or brooders. The turkeys are then turned out and in a few days will commence laying and will lay about as many eggs as they would have if not used as incubators. He thinks they are far superior to any artificial hatcher.

They are self-supporting, and a person is not obliged to fret or worry over severe changes of atmosphere, poor oil, unreliable thermometers, inferior mechanism, and the thousand and one other things that are so annoying to one running an incubator and especially a cheap incubator.

The truth is that artificial incubation is not an easy thing to accomplish. The incubator is not like the sewing machine or the corn sheller, which, if they are in good order, will do the work required of them as a matter of course, if a fair degree of mechanical skill is used in operating them. Hatching eggs artificially is a ticklish affair. It is a delicate operation to approach the particular required degree of heat very closely and yet never exceed it. Much money, much time, much labor is squandered in the pursuit of the fascinating employment of artificial hatching. It were well if novices would reflect that an old motherly hen is not to be sneezed at when incubation is desired. She can never get too hot. A feather is a most wonderful production. It is one of the most marvelous things in nature, securing, as it does, warmth, lightness, ventilation and all the combined requisites for both hatching the eggs and warming the tender younglings afterward.

FATTENING FOWLS.

If fowls are to be put upon the market, it is a prime requisite that they be suitably fat; a lean, scrawny bird will not be purchased from the stall as long as there is anything else to be found.

Unless the proper care is given, fowls will not become sufficiently fat for market without the outlay of more value in food, than the fowl would bring when sold. Beside, over-feeding, unless judiciously done, is a sure cause of disease; and similar effects follow too long or too close confinement.

To fatten fowls, shut from one to six in a movable coop, keep fresh water by them, give food three times a day, as early in the morning and as late at night as possible, giving all they will eat up clean. Use cooked food morning and noon. Indian meal, buckwheat, middlings, and mashed potatoes, with plenty of charcoal mixed into it. For supper, feed corn soaked in skim milk and a little wheat, and plenty of meat scraps with enough green food to keep their bowels open. Hang a piece of old carpet over the coop for an hour or two after each meal, as fowls fatten faster if kept dark and quiet. The coop should be moved daily that they may have fresh earth, but they should have as little exercise as possible to keep them in health. Any more than this calls for expenditure of food that avails nothing in the fattening process. They will not fatten, however, if restless and discontented. Their food should be given with great regularity and quietness. Some fowls will be nicely fattened at the end of five days; and in about fifteen days they begin to lose flesh again. All fowls should be cooped and fattened before killing, as the flesh is flavored by whatever the fowl has had for food.

POULTRY FOR MARKET.

All poultry intended for market should be well fattened, especially that sent for the holidays. The best manner of killing fowls is by bleeding in the neck; never wring the neck. Poultry intended for market should be dry picked, and if the feathers are plucked before the bodies are cold, this can be easily done. If poultry is scalded in the old fashioned way, it lessens the value fully-one third. After the fowl is dry picked, plunge it into a kettle of very hot water, holding it there long enough for the bird to plump; then hang it up, turkeys and chickens by the feet, ducks and geese by the head, until thoroughly cooled, This scalding makes the fat look bright and clear, and the fowl appears nice and plump. In packing, use clean packages, lining the sides and ends with paper, and cover over between the layers with clean rye straw. Pack as closely as possible so there will be no chance for the poultry to move about and become bruised. Good poultry will always sell for a full price, while common and inferior grades invariably sell low, and often prove a loss to the shipper. All poultry should be thoroughly cooled before packing, and if one wishes to use a little extra care, they can wrap each fowl before packing. This prevents dust and straw adhering to it, and adds much to the appearance. The box should have the name of the consignor;

the number and variety of contents; as well as the name of the consignee marked upon it.

The French mode of killing is excellent, as it causes instant death without disfigurement, and is done by opening the beak of the fowl, and with a sharp pointed and narrow bladed knife, making an incision at the back of the roof of the mouth, dividing the vertebra, and thus causing instant death.

In the Philadelphia, Baltimore and New York markets, as well as among the Paris and London dealers, chickens with white or light skins are preferred to those with yellow skins and consequently the Dorkings, Black Spanish, Houdans, Creves, and other white-skinned varieties or their crosses always bring the best prices, and are in the quickest demand, while in Boston and the other New England cities, and in Chicago, and perhaps some of the other large Western cities, where any decided preference has been expressed, the yellow skinned birds are in the greatest demand.

DRESSING FOWLS FOR DIFFERENT MARKETS.

Chicago.—Dry picked, heads off, legs on, entrails drawn.

Philadelphia.—Dry-picked, heads and legs on, entrails undrawn.

New York.—Undrawn, heads on, legs on, dry picked. They may, however, be either dry picked or scalded, the dry picked being one cent higher in price.

Boston.—Dry picked, heads off, legs on, entrails drawn.

When the heads are cut off the skin should be drawn over the neck and tied to prevent the poultry from getting bloody.

Some poultry men in killing, hold the fowl by the feet and strike it a sharp blow on the head with a stick about a foot long, and about the size of a broom handle, and then bleed in the mouth.

The best markets are New York, Philadelphia, Baltimore, Boston and Washington in the East, and Chicago and St. Louis in the West. But there is but little difference in prices between those places and some smaller cities.

The best months are March and April, but any time from January to June will bring good prices.

The best breeds for market are those with plump bodies and yellow skin and legs. The Plymouth Rocks, Brahmas, Wyandottes and Cochins are good. We may also include the Langshans, however, as well as the Houdans and Dorkings.

What is meant by "undrawn fowls" is that the heads, legs, feet and entrails must remain, the feathers only being removed.

THE CHICKEN'S ENEMIES.

CATS.

The greatest enemy known among town and village poultry yards is the *neighbor's cat*. Shot guns are good if properly applied. So, too, is strychnine, if judiciously placed. But as cats prowl by day as well as by night, it is an exceedingly hard matter to get them.

THE SKUNK.

The skunk visits the chicken yard principally on rainy or dark nights. As there are some good arguments against the shot gun or steel trap in this case, a box trap will be found best, as the skunk will not be offensive till hurt or frightened. This immersed in water till the animal is dead, is a safe and easy way of killing it. In a locality of skunks, the chicks should be shut up every night. A board set up in front of a coop will generally keep them out.

THE HAWK.

There are two kinds—the large hen-hawk, and the small pigeon hawk. If not disturbed, they will get to be very familiar, and never forget where the chickens ramble. The guinea fowl is rather noisy to have around, but it is generally understood that they will keep hawks at a distance, either by their cry or their instinct to fight the hawk.

Bits of bright tin suspended on poles or wires are also useful in frightening them away.

THE WEASEL.

A dead fowl under the roost in the morning, is the only evidence that a weasel has been there. It works by crawling on the roost, and tapping a vein and sucking the blood. Some writers recommend sprinkling oil of anise and strychnine on a piece of fresh and bloody meat.

THE FOX.

When chickens or turkeys wander a long way from home, or on back farms, the fox is the most troublesome. A good dog and gun are the best arguments to convince them that they are out of place.

THE RAT.

Rather the most annoying in villages, as they will hide under any old barn, shed, floor or wall, from which they will make their raids, day or night, and prey on the chickens till nearly as large as partridges. A cat which can be trusted, is an excellent exterminator. Poison, so set that other creatures cannot get it, is about as good as anything. Either the coop must be set at a distance from their hiding-places, or war must be waged on the rats.

THE OWL.

This bird always comes around in the night, and commonly takes chickens which roost out in trees. The owl always lights before it attacks a fowl. This suggests the steel trap or the housing of poultry, which is a preventative of the ravages of this *bird* of the night.

A GOOD CROSS IN POULTRY.

The Partridge Cochin, as many know, is a very large, compactly built, handsome fowl. The plumage of the cock is similar in some respects to the Brown Leghorn, and the penciling of the plumage of the Leghorn and Cochin hens are almost alike, but in every other respect, except combs and color of skin and legs, they are not to be compared with each other. The Partridge Cochin is an inveterate sitter and sticks to her nest closely and with great patience, while the Leghorn, except to deposit her eggs, has a supreme contempt for such business, and will not sit at all. The Cochin is so clumsy that a four foot pen keeps it comfortably confined, while the agile and light Leghorn scales the highest pickets and bids defiance to fence and enclosures. The Cochin is rather coarse-grained in flesh, while the Leghorn is delicate, and fine-boned. If the Leghorn is superior as an egg producer to the Cochin, the latter is heavier in feathering, and hardier. The cross of the Brown Leghorn cock with the Cochin hen, therefore, makes one of the best breeds known. All the good qualities of both are combined in one, and the result is a hen with medium carcass, good flesh, heavy feathering, and good laying qualities. The plumage is nearly the same as the Leghorn, the comb more erect, the legs not so heavily feathered as those of the Cochin, and such a fowl is active and good at foraging, as well as easily kept in confinement. The cross produces an exceedingly heavy fowl, and one that seems to lay better in winter than either of the originals from which it is derived, while they are even better for sitting than the Cochins, as they are somewhat lighter in weight. We commend such a cross to all.

HOW TO BREED EITHER SEX OF FOWLS.

To obtain the desired result, or predomination, of either sex, you have to mate your fowls as follows: For cockerels, you have to mate a one-year-old cockerel with hens not more than two years old, and you will get the desired results. For breeding pullets mate a three-year-old cock with one-year-old hens. I have never had any exception to this rule.

"To decide when eggs are fertile or not hold them between the thumb and forefinger in a horizontal position with a strong light in front of you. The unfertilized egg will have a clear appearance, both upper and lower sides being the same. The fertilized egg will have a clear appearance at the lower side, while the upper side will exhibit a dark or cloudy appearance.

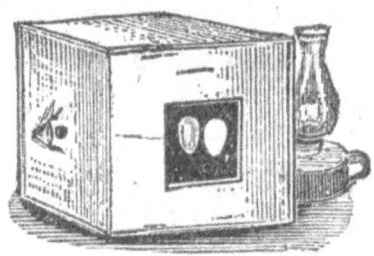

HOME MADE EGG TESTER.

Make a box eight or ten inches square, Cut holes as in illustration; then line box inside with anything dark to exclude light. The large hole on the side is to allow placing the hand inside so as to hold eggs. This tester can be used with the sun as well as a lamp.

Here is an article going the rounds of all the poultry papers that I do not exactly agree with. I quote it entire.

"Keep but one breed at first, of whatever kind you may fancy. When you can breed that well, try something else, if you get tired of this. But don't venture upon too much in the "variety line" at the commencement, or you will fail with all."

To be sure in making fences one need calculate on the kind of fowls kept, a three or four foot fence is all sufficient for the Asiatic, while a ten foot fence will no more than confine the light gamey breeds, Hamburgs, Leghorns and such. But in building a fence one might just as well put it at five or six feet, and then if found necessary, a wing can be clipped to confine the high-flying breeds. But I like a variety. There is a world of hen-nature if not human nature in all fowls,

and each breed have their own characteristics, wh
to me adds far greater interest and amusement to
care of fowls. So I would advise beginners to try
least two kinds, as light Brahmas and Houdans, Cochi
and Hamburgs, Plymouth Rocks and Leghorns, unles
of course, circumstances were not favorable.

THE ROOSTS.

A correspondent of the *Poultry Yard* follows the
plan described below, which, if carefully used, may be
safe and effective. He says: "I make saw-benches for
roosts—that is I take 2x4 scantling and nail legs to
them of the length to suit space and breed. These, as
you will at once see, can be moved at pleasure when
cleaning droppings or for any other reason that may
present itself; but the best part of it is as regards lice.
Every week I take my kerosene can and go to the hen-
house, carefully pick up and carry out of doors my saw-
benches, pour the kerosene on them and touch it off; the
flames will quicky run over every part of them, and go out
without burning the wood, but every mite, or nit that
would have made one, you are rid of forever. I have
followed this plan for some time, and have no trouble
with lice.

STOCK FOR BREEDING.

It is better when one can do so to purchase stock in
preference to eggs, as they will produce the eggs at
home in greater number than you can buy them for the
money the fowls cost, and you will be more likely to
have a *large* flock of chickens in the fall, beside the old
stock, to show for your money.

HOW TO PRESERVE FEATHERS.

The disposal and management of the feathers is a [th]ing that calls for attention. As soon as a fowl is [ki]lled, and while yet warm, let it be carefully plucked. [S]eparate the large wing-feathers; put the others into [s]mall paper bags previously prepared. Put these bags [i]nto an oven and let them remain about half an hour; take them out, repeat the process two or three times, then keep the feathers in a dry place till required. The oven must not be too hot. Care must be taken to free the feathers of any skin or flesh that may adhere to them while being plucked, or they will be tainted. The hard quilly portion of the larger feathers must be cut off with a pair of scissors. The wing and tail feathers may be stripped and added to the others. Previous to putting them in the oven, some recommend that the feathers should be put loosely into a dry tub or basket and shaken up daily, so that all may in turn be exposed to the air. Others recommend, as an easier plan, merely to suspend the bags from the ceiling of a warm kitchen or on the wall behind a fire-place, where it is practicable. In this case they will take longer to dry. Feathers can be quickly and effectually dried and cleaned by the agency of steam; but it is rather an expensive method, and the thrifty henwife will doubtless prefer having the produce of her own yard prepared under her own eye and by her own directions.

RECIPES.

CARBOLIC NEST EGG.

Open an egg shell at both ends, blow out the contents, fill with plaster of Paris, using a spoon to pour in the plaster, pour in carbolic acid slightly reduced with cold water, until the plaster will hold no more. In a few minutes the plaster will set, and you have a good, durable carbolic nest egg.

WATER-LIME PAINT.

Take lime-water—the stronger the better. Add salt to make quite a brine of it. Mix in water-lime to consistency of paint. If you can use half skim milk in place of half of the lime-water, it would add body and durability to the paint. Apply with whitewash brush.

FALSE NEST EGGS.

During all seasons of the year, and especially winter, a false nest egg is better than a real one. It is easily whittled from a large piece of chalk. Or take an egg, break a small hole in both ends, and blow the contents into a bowl, then fill the shell with plaster of Paris, and turn in water. It will harden, and be very solid.

CARBOLIC NEST EGGS.

Mix plaster of Paris in the form of an egg, and when dry, drop a little carbolic acid on it. This can be done whenever the egg loses its peculiar, acid odor. Such an egg will do much to keep the nests and hens free from vermin. It will not freeze or present any inducement for the hens to eat eggs.

SAND.

Sand is an important ingredient in poultry culture and is very cheap in most sections. Sand the poultry house floor every time it is cleaned out, and if it should require cleaning out, and the time to do it can not be spared, sprinkle the whole coop thickly with fresh sand. Sand the floors of the brood coops every day or two. Sand the filth board under the perches. Keep a pile of sand near the poultry house door, so no excuse will remain for neglecting it.

INDEX.

	PAGE.
Introduction	3
Poultry Houses and Yards	5
Plans for Houses	8
Early Chicken House	13
Coops	16
Yard and Fences	17
Hatching and Caring for Young Chicks	20
Artificial Heat	21
What to Feed	23
Incubators	26
Brooders	46
How to Preserve Eggs	53
Caponizing Fowls	61
Breeds of Fowls	65
Diseases of Fowls	130
Remedies	141
Miscellaneous Hints	151
Feed for Laying Hens	152
Feed for Poultry	154
To Fit Fowls for Exhibition	160
Packing Eggs for Hatching	164
Number of Fowls to the Acre	165
Poultry on Large Scale	170
The Turkey as an Incubator	172
Fattening Fowls	174
Poultry for Market	175
Dressing Fowls for Different Markets	177
The Chickens' Enemies	179
A Good Cross in Poultry	182
Home Made Egg Tester	184
How to Preserve Feathers	186
Miscellaneous Receipts	187

www.ingramcontent.com/pod-product-compliance
Lightning Source LLC
Chambersburg PA
CBHW062351220526
45472CB00008B/1771